LIBRARY-LRC
TEXAS HEART INSTITUTE

# Statistical Problems

*the text of this book is printed on 100% recycled paper*

# About The Author

A Cambridge, England, mathematician, Mr. L. H. Longley-Cook has made a career as an actuary in England and, since 1949, in the United States. He has written extensively on insurance statistics; their collection, interpretation, and use for determining insurance costs.

In 1968 he retired from the post of Vice-President-Actuary of the Insurance Company of North America and is now Special Lecturer and Research Consultant in the School of Business Administration at Georgia State University.

He is a Fellow of the Royal Statistical Society (England), the Institute of Actuaries, the Casualty Actuarial Society, and a member of other actuarial bodies.

His published books include a two volume actuarial text, *Life and Other Contingencies*, a booklet on *Credibility Theory*, and three mathematical puzzle books—*Work This One Out*, *Fun With Brain Puzzlers*, and *New Math's Puzzle Book*.

COLLEGE OUTLINE SERIES

# Statistical Problems

L. H. Longley-Cook

BARNES & NOBLE BOOKS
A DIVISION OF HARPER & ROW, PUBLISHERS
New York, Evanston, San Francisco, London

Reprinted, 1971



L.C. catalogue card number: 76-126340
SBN 389 00140 6

Manufactured in the United States of America

# Preface

The development of statistics can be divided into four principal periods. The first three were the early studies of political arithmetic of the seventeenth and eighteenth centuries; the development of frequency distributions in the nineteenth century; and the full exploration of mathematical statistics in the first half of the present century. Over the last two or three decades, statistics has entered its fourth and most exciting period. This has been the result of the spread of quantitative methods from the physical sciences to the social sciences, and to business, government, and the armed services.

It is now fully realized that every branch of the humanities needs statistical techniques to interpret its results. Biometrics has been followed by a whole series of other "—metrics" such as econometrics, sociometrics, etc., each applying statistical techniques to a specialized field. Further, the enormous developments in data processing equipment have enabled studies to be undertaken, particularly in the multivariate field, which were quite impractical a few years ago. Thus, statistics has become the vital tool of the research worker, from the physicist to the psychologist, from the archaeologist to the town-planner.

In addition, in the last two decades, there has been a vast spread of the applications of scientific methods to every type of organization and activity. This has led to a formalization of Decision Mathematics which brings together the techniques of not only statistics but those of applied mathematics, game theory, operations research, organization and methods, actuarial science, and many other fields. As a result, statistics is ceasing to be a specialized field of study but a part of basic education alongside formal mathematics.

Statistics is a very practical subject and it is the aim of this text, with its numerous examples, to provide the reader, not only with an understanding of statistical methods, but with the ability to apply these methods himself in whatever field of endeavor he is personally interested.

L. H. LONGLEY-COOK

*Georgia State University*

# Contents

# Statistical Problems

# 1

# Recording Data—Frequency Distributions

**1.1. Introduction.** We live in a world of uncertainty, and in every field of endeavor, decisions must be made in the face of uncertainty. In this highly competitive age, success is unlikely to be achieved by making guesses, relying on hunches, or even by judgment unsupported by the best data. In business, government, and military affairs, *decision mathematics* has emerged as the essential tool for making proper *judgment decisions*, and *statistics* is a major part of decision mathematics.

Statistics, originally consisting of the collection and tabulation of data concerning the *state*, is now the technical means by which data can be developed and analysed for *intelligent decision making*. The field of its use has extended far beyond state and government services, and it is employed extensively in every branch of science and business.

Because uncertainty is implicit in nearly everything we do, statistics is concerned with *probability distribution models*, *testing of hypotheses*, *significance tests*, and other means of determining the *correctness of our deductions* and the *most likely outcome* of our decisions.

The first step in the analysis of any source of information is the proper collection of statistics—the recording of data. The next step is the display of the recorded data in a form which can be readily interpreted. This chapter will be devoted to the study of these two problems. Statistics is a very practical subject, and a proper understanding of it can be best obtained by constant illustration in the text and the frequent working of examples. The text contains numerous worked examples, and each chapter is followed by a series of problems with, in most cases, detailed solutions.

It is often stated that you can use statistics to prove whatever you wish. This is not so, but the brazen use of statistics to mislead is all too common. "Six doctors out of ten recommend brand *X*." What does this mean? Were a large number of doctors asked which brand they thought best, or were they asked to list any number of different brands they might recommend? The interpretation of the statement depends greatly on the question posed. Perhaps just ten doctors were approached, and if less than six recommended brand *X*, another ten were tried. The statistician must look beyond the mere statistics he displays to interpret his results.

**1.2. Recording Data.** The first step in statistical analysis is the recording of data. It may be necessary to make a special study, or the data may be available from records maintained for a purpose other than the statistical investigation to be undertaken. In most cases data will be put on, or will be available on punched cards or tape. However, in the text, we show a limited volume of data in tabular form, similar to a write-out from a tabulating machine or a computer.

To illustrate the type of data which might be recorded, there is set out below, the first three lines of tabulated data concerning a group of students. The data have not been sorted in any order.

| Name | Sex | Age | Height | State of Birth | Number of Brothers and Sisters |
|---|---|---|---|---|---|
| Adams, John | M | 19 | 6 ft., 2 in. | N.Y. | 3 |
| Cowan, Tom | M | 22 | 6 ft., 0 in. | Pa. | 0 |
| Brown, Joan | F | 18 | 5 ft., 8 in. | N.Y. | 2 |

Such a tabulation is the "raw material" of any statistical investigation. It must be analysed and displayed if it is to be used to make reasonable deductions or comparisons with other similar data.

**1.3. Continuous and Discontinuous Variables.** It will be noted that some data are numerical and some are not. The name, sex, and the state are not numerical, although as will be explained later, numbers may be used for recording some *non-numerical data.*

Most of the data used in statistical work are numerical, and items recorded, such as age, height, and number of brothers and sisters, are each called *variables.* Some variables are discontinuous, or discrete in form, for example, the number of brothers and sisters, since the values which the variables can assume are limited to 0, 1, 2, 3, etc. (The term "half brother" does not, of course, have the same meaning as "half a brother." Instructions for any tabulation must explain whether or not half brothers are to be included as brothers.)

While age and height are *continuous variables*, they and other continuous variables have to be recorded in discrete form. Height is usually expressed in feet and inches; it could be measured in some other unit such as centimeters.

**1.4. Processing Data.** The days of hand tabulation, hand sorting, and calculation on simple calculating machines are a thing of the past for any sizeable investigation. The vast majority of statistical work is carried out by means of electronic data processing equipment, or by punched cards.

For simplicity of recording data when punched cards are used, and to conserve space on the card, non-numerical data may be recorded in an alphabetical or numerical code. Thus, *M* would be used for Male and *F* for Female, or where cards are used, sex and marital status might be indicated in a single code:

1. Male, Single
2. Male, Married
3. Female, Single
4. Female, Married

Similarly, two digits can be used to record states by code.

Although it is rarely necessary to sort data by hand, some understanding of how this can best be done is necessary, and this is explained in the next section. Similarly, short cuts which can be used in the calculation of statistical constants by hand will be included in the text, although such short cuts are of considerably less importance now that most work is done by processing equipment.

In order to study a limited volume of data, the raw data are often arranged in an *array* of descending order of magnitude. The difference between the largest and smallest value is called the *range* of the data.

EXAMPLE 1.1. The ages of the sixteen members of a bridge club are:

37, 28, 40, 47, 30, 42, 38, 45,
47, 39, 52, 25, 30, 35, 29, 41

Arrange these data in an array of ascending age and determine the range of the data.
*Solution.* The array is:

25, 28, 29, 30, 30, 35, 37, 38,
39, 40, 41, 42, 45, 47, 47, 52

and the range is:

$$52 - 25 = 27 \text{ years}$$

Occasionally, it is necessary to obtain an array for a large volume of data without the help of data processing equipment. In such a case, it is best to select suitable groups of ages, and tabulate for each group the ages of the individual persons in the group. For the above data the tabulation would be as follows:

| Age Group | Individual Ages |
|---|---|
| 20–29 | 28, 25, 29 |
| 30–39 | 37, 30, 38, 39, 30, 35 |
| 40–49 | 40, 47, 42, 45, 47, 41 |
| 50–59 | 52 |

The final array can then be written down without difficulty.

25, 28, 29, 30, 30, 35, 37, 38,
39, 40, 41, 42, 45, 47, 47, 52

**1.5. Use of a Score Sheet to Tally Data.** With a larger volume of data, there will be a number of persons with the same age, and instead of an array, it is useful to tabulate the *number* of persons at each age. Similar tabulations can be made for other numerical data.

When data are being analyzed without the help of data processing equipment, use is made of a *score sheet* to *tally* the age or other measure of the data. The next example shows how this is done.

EXAMPLE 1.2. The ages of the students in a class are:

19, 20, 23, 25, 21, 19, 19, 18, 20, 20,
17, 23, 19, 21, 22, 18, 20, 19, 20, 21,
21, 18, 20, 19, 21, 20, 17, 18, 22, 19,
22, 20, 19, 22, 20, 21, 19, 20, 19, 18,
22, 18, 22, 20, 22, 19, 20, 23, 22, 26,
21, 19, 20.

Use a score sheet to find the frequency distribution of students by age and determine the range of ages.
*Solution.*

| Age | Tally | Total Number |
|---|---|---|
| 17 | \|\| | 2 |
| 18 | ~~\|\|\|\|~~ \| | 6 |
| 19 | ~~\|\|\|\|~~ ~~\|\|\|\|~~ \|\| | 12 |
| 20 | ~~\|\|\|\|~~ ~~\|\|\|\|~~ \|\|\| | 13 |
| 21 | ~~\|\|\|\|~~ \|\| | 7 |
| 22 | ~~\|\|\|\|~~ \|\|\| | 8 |
| 23 | \|\|\| | 3 |
| 24 | | 0 |
| 25 | \| | 1 |
| 26 | \| | 1 |
| | | 53 |

It will be noted that every fifth tally is drawn on the diagonal across the four preceding tallies to assist in counting. The range of the data is

$$26 - 17 = 9 \text{ years}$$

**1.6 Class Intervals.** In statistical work, a major problem is how to reduce a large volume of raw data into a form in which the chief characteristics of the data can be studied and comparisons made with other related data. For this reason, the use of arrays for large volumes of data is inappropriate, and frequency distributions are developed which show the distribution of data, not by individual values of such measures as ages, heights, or examination scores, but by ranges of such values. Thus, the tabulating of the residents of a town by age might use age groups 0–5, 6–10, 11–15, 16–20, 21–25, etc., or 0–9, 10–19, 20–29, etc. Such groups are called *classes*. It is important that classes should not overlap and that there should be no gaps between classes. Class intervals need not be of equal size but the graphical display of data and the calculation of statistical constants becomes slightly more difficult when unequal intervals are used.

If height of trees is measured in feet, the class intervals could be 0–9 feet, 10–19 feet, etc., but if height is measured in feet and inches, the class intervals would be 0 feet–9 feet, 11 inches; 10 feet–19 feet, 11 inches, etc. In either case the class intervals can be expressed as, 0 feet–, 10 feet–, 20 feet–, etc.; only the commencing point of each interval being stated, followed by a dash.

When data processing equipment is used, class intervals 0–9 feet, 10–19 feet, etc., are more satisfactory than 1–10 feet, 11–20 feet, etc., because the sorting machines can be instructed to sort by the first digit only in the former case, but not in the latter.

With height measured in feet, the class interval 10 feet to 19 feet includes all true measurements from $9\frac{1}{2}$ feet to $19\frac{1}{2}$ feet, since 9.51 feet would be rounded to 10 feet. $9\frac{1}{2}$ feet is called the *lower class boundary** and $19\frac{1}{2}$ feet the *upper class boundary*. The difference between the upper and the lower class boundary is called the *class size*, width, or interval. In the example being used, the class size is

$$19\tfrac{1}{2} \text{ feet} - 9\tfrac{1}{2} \text{ feet} = 10 \text{ feet}$$

The midpoint of a class interval is important, and is obtained by adding the lower and upper class boundaries and dividing by 2. In the example used above, the midpoint is

$$\frac{19\frac{1}{2} + 9\frac{1}{2}}{2} = 14\tfrac{1}{2} \text{ feet}$$

This is called the *class midpoint* or *class mark*.

It is the custom to record ages as *age last birthday*; hence, the class interval, age 20 to age 29, will include all persons with age last birthday 20 to 29 inclusive. The class boundaries will be *exact* age 20 and *exact* age 30. The class size will be 10 years and the class midpoint will be *exact* age 25.

Class intervals need not commence from zero but must include the lowest and highest values. To choose class intervals, determine the highest and lowest values, and hence, determine the

*In this example, 10 feet is sometimes referred to as the lower class *limit*, and 19 feet as the upper class *limit*.

range. If possible, the class intervals should be chosen by dividing the range into a convenient number (about 10 is often suitable) of equal intervals. However, this may not always be possible. The midpoints of class intervals should, if practical, coincide with actual observed data.

EXAMPLE 1.3. If the data in Example 1.2 is to be displayed in three classes, what class intervals should be used?
*Solution.* With 10 ages, equal class intervals must include more than three ages each and the most suitable intervals are

16–19
20–23
24–27

EXAMPLE 1.4. The results of an examination were as follows:

| Marks | Students |
|---|---|
| 40–49 | 1 |
| 50–59 | 7 |
| 60–69 | 23 |
| 70–79 | 21 |
| 80–89 | 8 |
| 90–99 | 5 |
| | 65 |

Determine the following:

| | *Solution* |
|---|---|
| 1. The number of classes | 6 |
| 2. The class size | 10 |
| 3. The lower class boundary of the third class | $59\frac{1}{2}$ |
| 4. The class midpoint (or mark) of the fifth class | $84\frac{1}{2}$ |

EXAMPLE 1.5. If class intervals for a certain measurement are 0–, 5 inches –, 10 inches –, 15 inches –, and measurements are made to the nearest 1/8 of an inch, what are the class boundaries, size, and midpoint of the second class?
*Solution.*

| | |
|---|---|
| Lower class boundary | $4\frac{15}{16}$ inches |
| Upper class boundary | $9\frac{15}{16}$ inches |
| Class size | 5 inches |
| Class midpoint | $7\frac{7}{16}$ inches |

**1.7. Rounding Numbers.** Data, whether in the form of a continuous or discontinuous variable, may have been rounded before they reach the statistician, but the statistician may often desire to round numbers to reduce the size of the figures to be tabulated. Thus, the population of the United States at the 1960 Census was 183,285,009. For normal use, no lack of accuracy would result from rounding the figure to the nearest thousand—183,285,000, or even to the nearest hundred thousand—183,300,000. It will be noted that in the latter example, 183,300,000 is the *nearest* hundred thousand, so as to make the approximation as near the true figure as possible. This result is better written as $1.833 \times 10^8$ or 183.3 million. The student should avoid recording figures or calculating averages and other statistical constants, to a greater number of digits than are meaningful. Thus, in calculating the average height of the students in a class, 5 ft., 11 in. is a meaningful figure but 5 ft., 11.23 in. includes two places of decimals which are meaningless, since the height of an individual cannot be measured to a tenth or a hundredth of an inch.

The rounding of a number such as 17.50 to the nearest integer presents a dilemma because 17.50 is equally distant from 17 and 18. One practice is to round alternate cases up and down so as to minimize the cumulative rounding error. However, with the increased use of data processing equipment, it is more usual to "round up," to 18 in this case, because this lends itself more readily to machine processing.

EXAMPLE 1.6. Round the following numbers to the nearest unit, 10 units, and 100 units. 47.73, $392\frac{3}{4}$, $72\frac{1}{8}$, 5321.09, 400, 74.7, 155, 149.5.

*Solution.*

| | | | | | | | | |
|---|---|---|---|---|---|---|---|---|
| **Units** | 48 | 393 | 72 | 5321 | 400 | 75 | 155 | 150 |
| **10 Units** | 50 | 390 | 70 | 5320 | 400 | 70 | 160 | 150 |
| **100 Units** | 0 | 400 | 100 | 5300 | 400 | 100 | 200 | 100 |

**1.8. Frequency Distributions.** While many types of descriptive statistical distributions are possible, such as distributions according to sex, location, plant genus, color, etc., this book will be concerned mainly with two special types of distribution which have particular importance. These are *frequency distributions* and *time series.* Frequency distributions are discussed in this and in the following chapters. Time series are considered in Chapter 9.

The *frequency* of a variable $x$ is the number of times it occurs, and a *frequency distribution* is an arrangement of numerical data which displays the frequency of the data according to a variable which measures *size or magnitude.* Example 1.2 shows the frequency distribution of a class of 53 students according to age. Example 1.4 shows the frequency distribution of a class of 65 students according to examination marks. It will be noted, as is usually the case, that the marks are grouped into class intervals and the *frequency of each class interval* is shown. Frequency distributions can be displayed in graphical form as in Figures 1.1 and 1.2. In this case the independent variable is measured along the horizontal, or $x$ axis, and the number or frequency, the dependent variable, is measured along the vertical, or $y$ axis.

It should be noted that the class interval grouping is a measure of *size or magnitude* such as age, height, weight, angle, distance, etc. It is not really appropriate to use the term frequency distribution for a non-numerical class measure such as a tabulation of the distribution of population by states, although this term is sometimes used in such cases.

A *frequency histogram* consists of a series of rectangles with bases on the horizontal ($x$) axis with centers at the class midpoints and length equal to the class interval. The *areas* of the rectangles are proportionate to the class frequencies. When the class intervals are equal in size, the *heights* of the rectangles are also proportionate to the class frequency.

A *frequency polygon* is obtained by joining the midpoint of the top of each rectangle. When the class intervals are equal, it is a graph of class frequency against class midpoint.

In order to compare two frequency distributions, it is useful to increase or decrease the class frequencies so that the total of all class frequencies is the same for both distributions. Such results are usually expressed as percentages of the total for all classes and are called *relative or normalized frequency distributions.*

EXAMPLE 1.7. Draw a frequency histogram and frequency polygon for the data in Example 1.2 and calculate the relative frequency distribution.

Figure 1.1. Age distribution of a class of students. Frequency histogram and frequency polygon.

*Solution.* The histogram and polygon are shown in Figure 1.1. The relative frequency distribution is obtained by dividing the total in each class by 53, the total number of students.

| Age* (1) | Number of Students (2) | Relative Frequency Col. (2) ÷ 53 |
|---|---|---|
| 17 | 2 | 4% |
| 18 | 6 | 11% |
| 19 | 12 | 23% |
| 20 | 13 | 25% |
| 21 | 7 | 13% |
| 22 | 8 | 15% |
| 23 | 3 | 6% |
| 24 | 0 | 0% |
| 25 | 1 | 2% |
| 26 | 1 | 2% |
| | 53 | 101% |

*Last birthday

The total percentage is 101%, not 100%, because of the rounding of the individual percentages to the nearest one percent.

EXAMPLE 1.8. The following grouped data show the results of an end of term examination. Draw the frequency histogram and frequency polygon of the results.

| Grade | Number of Students |
|---|---|
| Under 30 | 6 |
| 30–39 | 4 |
| 40–49 | 7 |
| 50–59 | 10 |
| 60–69 | 20 |
| 70–79 | 18 |
| 80–99 | 5 |
| 100 | 0 |
| | 70 |

*Solution.* This is a case of class intervals which are not all the same size, the first and next to last groups being larger than the others. Care must be used in such cases. When the class interval is twice the normal, as in the next to last interval, the length of the rectangle will be twice normal, and hence, the height must be one-half the number of students in order to make the histogram area proportional to the students. The first interval is three times the normal so we make the height one-third the number of students in the class, or 2. This is the same as assuming:

| Grade | Number of Students |
|---|---|
| 0–9 | 2 |
| 10–19 | 2 |
| 20–29 | 2 |
| 0–29 | 6 |

The histogram and polygon are shown in Figure 1.2.

Remember that for the class 30–39, for example, the lower and upper boundaries are $29\frac{1}{2}$ and $39\frac{1}{2}$ and the midpoint is $34\frac{1}{2}$. The divisions between the rectangles are drawn at these boundaries and the points on the polygon are at the midpoints of the classes.

**1.9. Cumulative Frequency Distribution.** Sometimes it is desirable to show the distribution of data in a cumulative form. Thus, we may wish to tabulate the number of students scoring less than or more than various scores in the example above. These distributions are called *cumulative frequency distributions* or *ogives* and are obtained by summing the frequency distribution from the top or from the bottom of the table. As with frequency distributions, it is often useful for comparative purposes to develop *relative*

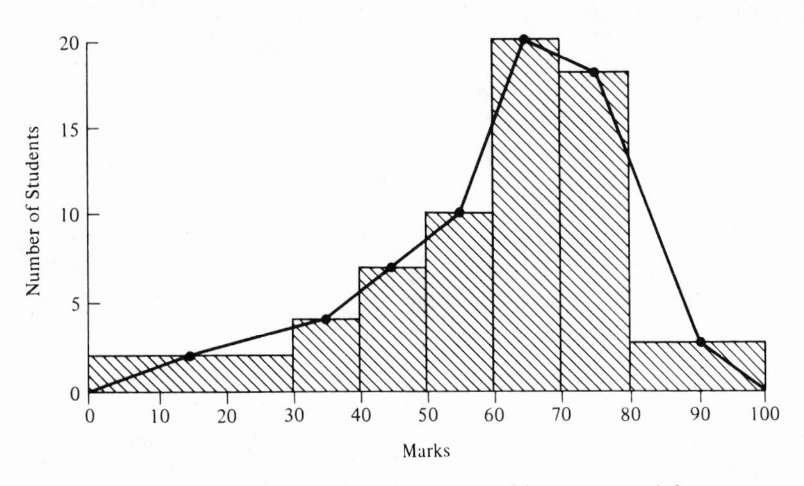


Figure 1.2. Examination results. Frequency histogram and frequency polygon.

*cumulative frequency distributions* or *percentage ogives* by expressing the cumulative frequencies as a percentage of the total frequency.

EXAMPLE 1.9. Tabulate "less than" and "and over" ogives for the data in Example 1.8 and graph the results.
*Solution.*

| Marks | Number of Students |
|---|---|
| Less than 30 | 6 |
| Less than 40 | 10 |
| Less than 50 | 17 |
| Less than 60 | 27 |
| Less than 70 | 47 |
| Less than 80 | 65 |
| 100 or less | 70 |

| Marks | Number of Students |
|---|---|
| 0 and over | 70 |
| 30 and over | 64 |
| 40 and over | 60 |
| 50 and over | 53 |
| 60 and over | 43 |
| 70 and over | 23 |
| 80 and over | 5 |
| Over 100 | 0 |

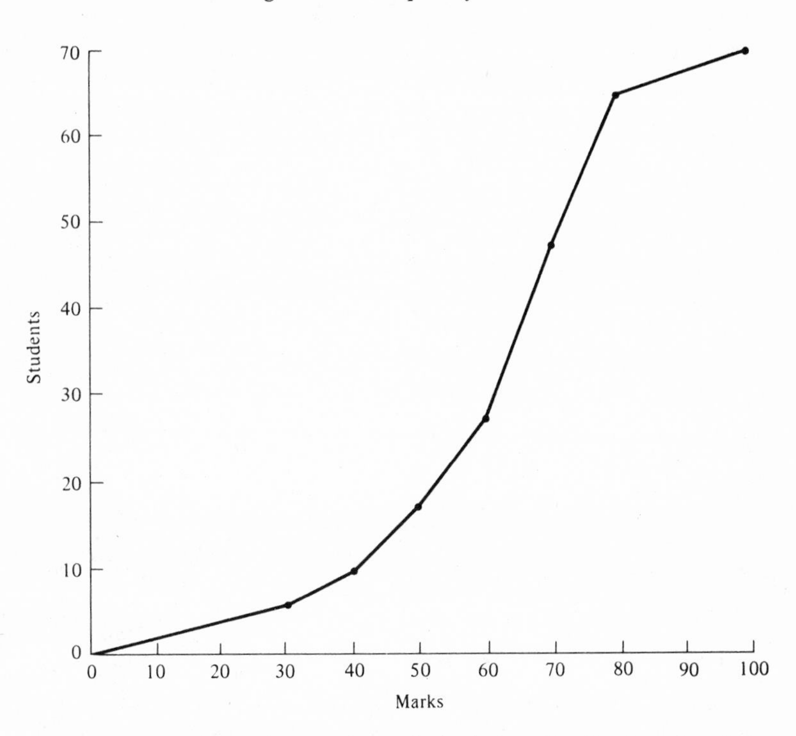


Figure 1.3. Examination results. Cumulative distribution function. "Less than" ogive of the data in Figure 1.2.

The graphs of these distributions (ogives) are shown in Figures 1.3 and 1.4.

**1.10. Frequency Curves.** In most statistical studies, the data are only a *sample* of a larger *universe* of data. If the whole universe of data were examined, very small class intervals could be used and it would be generally found that the frequency polygon approximates a smooth curve. It is convenient, therefore, to draw a smooth curve which follows the indication of the frequency polygon but does not necessarily pass through any of the actual points of the polygon. Such a curve is called a *frequency curve*. In the same way a smooth ogive can be drawn to fit a cumulative frequency distribution.

Three of the most common shapes of frequency curves are illustrated in the top half of Figure 1.5 and are called *bell-shaped*. Three less usual shapes are shown in the bottom half. However, curves with more than one maximum value can occur.

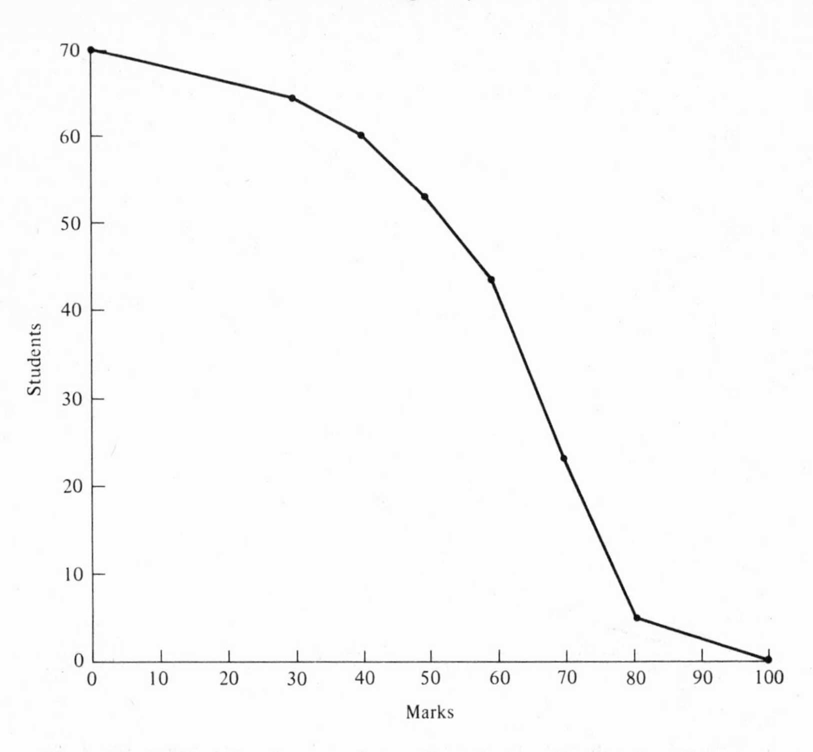


Figure 1.4. Examination results. Cumulative distribution function. "And over" ogive of the data in Figure 1.2.

It will be noted that while the first curve in the top row is symmetrical, the second curve has a "tail" which measured from the highest point, is longer on the right than on the left. Such a curve is stated to be skewed to the right. A discussion of skewness will be found in Section 5.5 of Chapter 5 where the significance of positive (+ ive) and negative (– ive) skewness will be explained.

EXAMPLE 1.10. Draw a frequency curve for the distribution by age of the students in a class given in Example 1.2.

*Solution.* The center points of the top of each rectangle of the frequency histogram are the points of the frequency polygon. These are indicated by dots on graph paper and a smooth curve is drawn passing through these points, keeping the dots as equidistant as possible on either side of the curve, as illustrated in Figure 1.6.

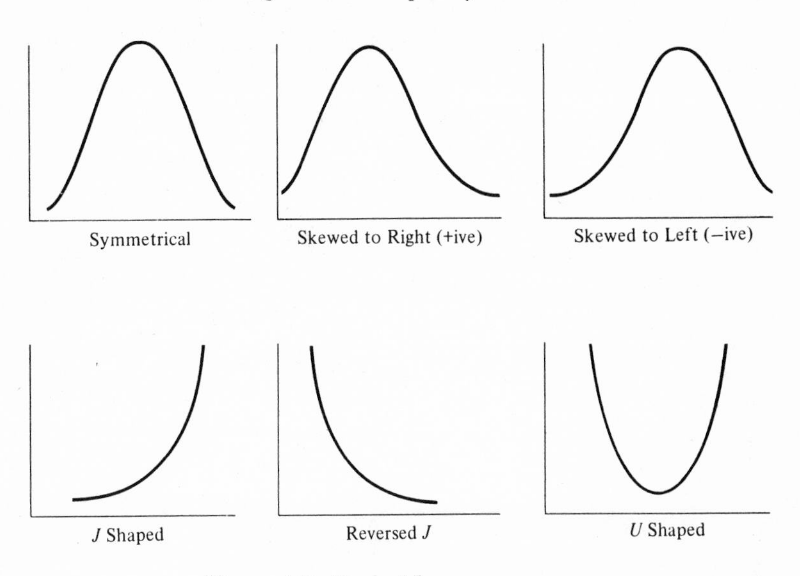


Figure 1.5. Typical frequency curves.

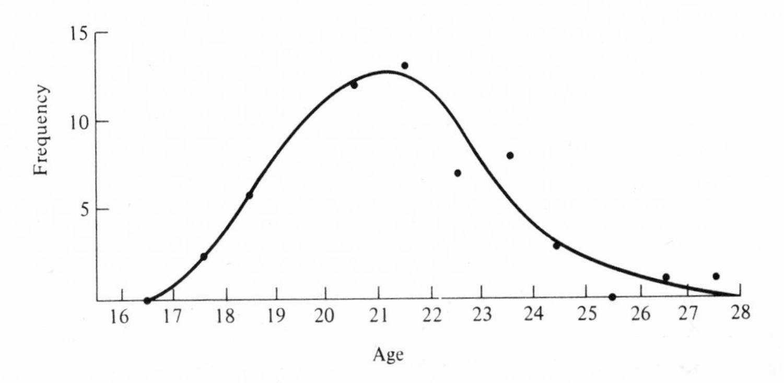


Figure 1.6. Age distribution of a class of students. Smooth frequency curve.

Example 1.11. As part of a study of variations in the earth's magnetism over long periods of time, the magnetic inclination of a number of lava flows from volcanic formations in Hawaii were analyzed and correlated with the age of the flows. The following table gives the frequency distribution of these flows by inclination angle. Draw a frequency histogram and a frequency curve of the data.

| Inclination Angle | Number of Lava Flows |
|---|---|
| −10°– | 2 |
| 0°– | 6 |
| 10°– | 18 |
| 20°– | 27 |
| 30°– | 26 |
| 40°– | 21 |
| 50°– | 7 |
| 60°– | 0 |
| | 107 |

*Solution.* The frequency histogram is given in Figure 1.7, and the frequency curve in Figure 1.8. The dots on the frequency curve are the points of the frequency polygon.

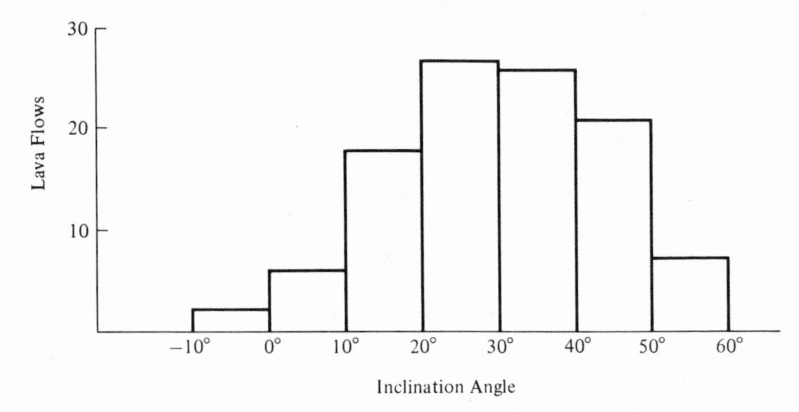


Figure 1.7. Lava flows in Hawaii. Frequency histogram.

## Problems

**Problem 1.1.** List six continuous and six discontinuous variables, not mentioned in the text, which might be used in statistical work.

**Problem 1.2.** What is the minimum number of digit codes (0 to 9 inclusive) which are needed to record:

*a*. Two "yes or no" answers.
*b*. Three "yes or no" answers.
*c*. One two-choice answer and one three-choice answer.
*d*. The day of the month.
*e*. The month of the year.
*f*. The year.

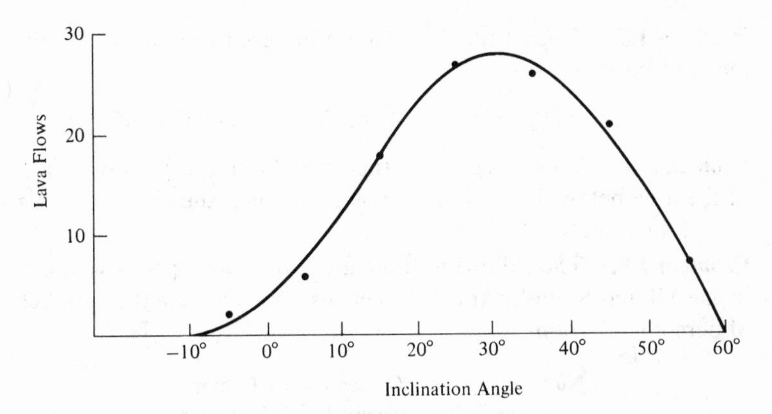


Figure 1.8. Lava flows in Hawaii. Frequency curve.

**Problem 1.3.** Use a tally score sheet to determine the frequency of the number of words on a line of a page of print. Before you start, write down the shape of the frequency curve you expect to get.

**Problem 1.4.** By means of a tally sheet, obtain the frequency of the letters of the alphabet in the wording of this sentence. Why do these data not meet the definition of a frequency *distribution* used in this book?

**Problem 1.5.** Suggest suitable class intervals for a study of the height of male college students when a maximum of nine codes can be used.

**Problem 1.6.** The following is part of a tabulation of the residents of Pennsylvania in 1960, analyzed according to adjusted gross income:

| Adjusted Gross Income Class | Number of Returns |
|---|---|
| Under $1,000 | 79,120 |
| $1,000–$1,999 | 292,734 |
| $2,000–$2,999 | 376,775 |
| etc. | etc. |

For the second class, what are:

(1) the class limits
(2) the class boundaries
(3) the class size
(4) the class midpoint?

**Problem 1.7.** Round the following numbers to the nearest unit, ten, and hundred.

541.7, $17\frac{1}{2}$, 155.5, 4545.4, 957, 62.47, 3/4, 85

**Problem 1.8.** Which is greater, the area of a frequency histogram or the area below the frequency polygon, when the class intervals are all the same size?

**Problem 1.9.** The following table gives the number of dwellings in the United States at the 1960 census, analyzed by the number of persons per room.

| Number of Persons Per Room | Owner Occupied Dwellings (In thousands) | Renter Occupied Dwellings (In thousands) |
|---|---|---|
| 0.50 or less | 15,292 | 6,897 |
| 0.51 to 0.75 | 7,702 | 4,756 |
| 0.76 to 1.00 | 6,953 | 5,310 |
| 1.01 to 1.50 | 2,180 | 2,031 |
| 1.51 or more | 670 | 1,233 |
| | 32,797 | 20,227 |

Construct a frequency histogram and a frequency polygon of the distribution of owner occupied dwellings by number of persons per room. Draw a frequency curve which fits these data. (Note that the class intervals are not all the same size and an assumption must be made as to the upper limit of the last class.)

**Problem 1.10.** Using the data of Problem 1.9, develop a frequency distribution for the renter occupied dwellings and for all dwellings.

**Problem 1.11.** Tabulate the "less than" cumulative distribution functions for the data in Problem 1.9, for owner occupied, for renter occupied, and for all dwellings, and set out the results in graphic form.

**Problem 1.12.** Tabulate and graph the "more than" cumulative distribution functions for the data in Problem 1.9 in the same manner as in Problem 1.11.

**Problem 1.13.** Take a pack of shuffled playing cards and deal out successive groups of three cards and tally the resultant frequencies:

| Maximum Number of Cards in Any Suit | Tally | Total |
|---|---|---|
| 1 (i.e., all cards of different suits) | | |
| 2 (i.e., two cards of the same suit) | | |
| 3 (i.e., three cards of the same suit) | | |

Calculate the relative frequency distribution and draw the frequency polygon. (Try to forecast which group will have the highest frequency before starting the tally.)

**Problem 1.14.** In a census of the population of the United States, the number of persons at each age is recorded. If a graph of this frequency distribution were drawn, what would be its shape?

**Problem 1.15.** Suggest statistical distributions which are likely to be:

(*a*) Bell-shaped—Symmetrical
(*b*) Bell-shaped—Skewed to right
(*c*) Bell-shaped—Skewed to left
(*d*) *J* shaped
(*e*) Reversed *J*
(*f*) *U* shaped

**Problem 1.16.** The following table gives the live births per 1000 native white women, aged 15–44, in the United States in 1964:

| Birth | Number |
|---|---|
| First | 30 |
| Second | 20 |
| Third | 11 |
| Fourth | 6 |
| Fifth | 4 |
| Sixth | 2 |
| Seventh | 2 |
| Eighth and Over | 3 |

Draw a frequency polynomial. What is the type of this frequency distribution?

**Problem 1.17.** Calculate and draw an "or over" ogive from the data in Problem 1.16. Calculate the percentage ogive for this data.

**Problem 1.18.** Six coins are tossed and the number of heads counted. The result may be any number from 0 to 6. The experiment is performed 100 times, and the results are classified as follows:

| **Heads** | 0 | 1 | 2 | 3 | 4 | 5 | 6 |
|---|---|---|---|---|---|---|---|
| **Number Observed** | 2 | 7 | 26 | 29 | 24 | 10 | 2 |

Draw a graph of this frequency polygon. Has a smooth frequency curve any significance in this case?

**Problem 1.19.** A true die is rolled 100 times and a tally made of the number of times 1, 2, 3, 4, 5, or 6 comes up. What would be the shape of the frequency distribution?

**Problem 1.20.** Role two dice 100 times and tally the results. Draw a graph of the frequency polygon. What shape would you expect this polygon to have?

## Solutions

**Problem 1.1.** Continuous: weight, girth, speed, elevation, illumination, miles of highway, etc. Discontinuous: *Number* of very many things such as houses, sales, employees, radios, rooms in a house, children, etc.

**Problem 1.2.** (*a*) 1, (*b*) 1, (*c*) 1, (*d*) 2, (*e*) 2, (*f*) 2*

*Rarely will there be the need to record more than the last two digits of the year (e.g., 67 for 1967) and in many statistical studies a single digit is sufficient.

**Problem 1.3.** If there are a full number of words on each line, the frequency curve will be bell shaped, but if short lines in headings and at the end of paragraphs are included in the study, the frequency curve will be skewed to the left.

**Problem 1.4.** The score sheet will start as follows:

| **Letter** | **Tally** | **Total Number** |
|---|---|---|
| A | 𝍸 \| | 6 |
| B | \|\|\| | 3 |
| C | \|\| | 2 |
| D | \| | 1 |
| E | 𝍸 𝍸 𝍸 \| | 16 |
| etc. | | |

Since the letters A, B, C, etc., are not measures of size or magnitude, this is not a frequency *distribution* as defined in Section 1.8. Considerable use can be made of such distributions as

# 2

# The Mean of a Distribution

**2.1. Definition.** Much statistical data consists of a tabulation of the value of some characteristic for each of a number of individuals or items, for example, the ages of students in a class. The most obvious measure to describe this characteristic of the data is an average value of the variable; in this case the average age of the class. While various averages may be used, the one most frequently employed in statistical work is the *arithmetic mean* or *mean* as it is usually called. This is the arithmetic average obtained by adding up all the individual values and dividing by the number of items. There are other means, such as the geometric mean, described in Chapter 3, but when the term *mean* is used without qualification the arithmetic mean is meant. The mean provides a single value which is typical of the data and is an indication of the *point of central tendency*.

**2.2. Calculation of the Mean.** With data recorded on punched cards or tape as is usually the case, the calculation of the mean is very simple and straightforward. Data processing equipment can be readily instructed to add all the ages or other characteristic which is being studied. All that remains to be done to calculate the mean is to divide the total by the number of students (or other base as appropriate). With more sophisticated equipment even the division can be performed automatically. However, the student must learn how to calculate the mean of a distribution without the advantage of data processing equipment. Various procedures are available which reduce the work of handling large volumes of data and these procedures are explained in the following examples.





this one, particularly when the data are rearranged so that the letter (or other classification such as country, sales outlet, etc.) with the greatest frequency is placed first, the letter with the next greatest frequency is placed second, etc. If we do this, we find the order of letters is:

E, T, N, H, A, O,...

It is interesting to note how close this order, which has been obtained for a small sample, is to the order established for English writing from extensive studies. This order is:

E, T, A, I, S, O, N, . . .

**Problem 1.5.** It is unlikely that a student will be under 4′6″ or over 6′9″, although an occasional rare case can occur. Suggested class intervals are:

4′6″–, 4′9″–, 5′0″–, 5′3″–, 5′6″–,
5′9″–, 6′0″–, 6′3″–, 6′6″–.

**Problem 1.6.** The answers are:

(1) \$1,000 and \$1,999
(2) \$999.50 and \$1,999.50
(3) \$1,000
(4) \$1,499.50

However, the statistics collected will not have sufficient accuracy to justify this exactitude and the class boundaries can be taken as \$1,000 and \$2,000 and the class midpoint as \$1,500 without loss of accuracy.

**Problem 1.7.**

| | | | | | | | | |
|---|---|---|---|---|---|---|---|---|
| **Units** | 542 | 18 | 156 | 4545 | 957 | 62 | 1 | 85 |
| **Tens** | 540 | 20 | 160 | 4550 | 960 | 60 | 0 | 90 |
| **Hundreds** | 500 | 0 | 200 | 4500 | 1000 | 100 | 0 | 100 |

**Problem 1.8.** Consider two neighboring class rectangles *ABCD* and *EFGC* in Figure 1.9. The midpoints of the tops of these rectangles are *X* and *Y*. The portion of the histogram excluded from the area below the polygon is the triangle *EYZ* and this triangle is exactly equal to the triangle *BXZ*, the portion of the area below the polygon excluded from the histogram. Hence, the two areas are equal.

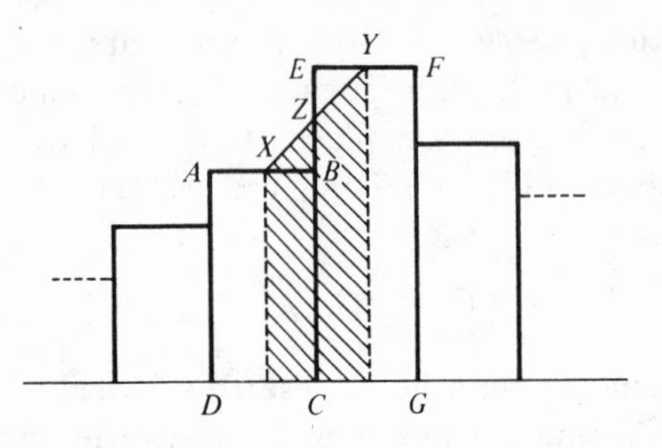


Figure 1.9. Frequency histogram and frequency polygon. Comparison of areas.

**Problems 1.9–1.12.** These are straightforward and no solutions provided.
**Problem 1.13.** Class 2 (two cards the same suit) is likely to have the highest frequency.
**Problem 1.14.** The number of persons at each attained age will generally decrease as the age increases, but owing to fluctuations in the number of births each year, this rule will not be exact and there will be cases where the number of persons at age $x + 1$ is greater than the number of persons at age $x$.
**Problem 1.15.**

(*a*) Symmetrical bell-shaped distributions arise from many pure probability concepts such as occur when rolling dice or tossing pennies. Also, many studies in biology (such as distributions by size) will develop curves very close to bell shaped.

(*b*) Bell-shaped distributions skewed to the right occur frequently in studies of wealth, such as distributions of income, size of houses, etc.

(*c*) Bell-shaped distributions skewed to the left are likely to occur in distributions of examination marks, I.Q. scores, etc.

(*d*) *J* shaped distributions will occur only when there is a top limit to the class interval. An example would be the number of games in a world series where the maximum of seven games will be more frequent than the minimum of four games.

(*e*) The number of automobile drivers having 0, 1, 2, 3 or more accidents a year provides an example of a reversed *J* curve.

(*f*) *U* shaped distributions, like *J* distributions, can occur only when there is a top limit to the class interval. An example would be deaths per 100 persons living, classified according to age. The



rate of mortality in the first year after birth is higher than in sequent years, and after about age 25, the rate of mortality creases with each year of age.
**Problems 1.16 and 1.17.** Straightforward; no solutions provic
**Problem 1.18.** Because of the discrete nature of the independ variable (the number of heads), a value corresponding to heads has no significance. However, a smooth curve does ena the eye to get a clearer picture of the distribution.
**Problem 1.19.** Since any number from 1 to 6 is equally likely the roll of a die, the frequency of each score would be appro mately 17 and a frequency polygon would approximate to straight line.
**Problem 1.20.** A symmetrical bell-shaped distribution.

EXAMPLE 2.1. Calculate the mean I.Q. (Intelligent Quotient) of a group of eight students with the following I.Q.'s: 100, 105, 95, 110, 100, 85, 95, 95.
*Solution.* In this simple example as there are only a few values the direct procedure of adding up the scores can be used.

$$100 + 105 + 95 + 110 + 100 + 85 + 95 + 95 = 785$$

The mean is then obtained by dividing by 8, giving a mean I.Q. of 98. Note that the actual mean is $98\frac{1}{8}$ but when the I.Q.'s of individual students are only given in numbers which are rounded to multiples of 5, anything beyond two significant figures in the result is meaningless.

EXAMPLE 2.2. The number of rooms in owner occupied dwellings in a certain district are available in the form of a listing 5, 3, 8, 4, 5, .... How would you proceed to obtain the mean number of rooms per dwelling?
*Solution.* The first step is to make a score sheet as illustrated in Chapter 1, Section 1.5. Assumed totals have been included to make the procedure clear.

| Number of Rooms in Owner Occupied Dwellings (1) | Number of Dwellings (2) | Total Number of Rooms in Class Col. (1) × Col. (2) (3) |
|---|---|---|
| 1 | 1 | 1 |
| 2 | 4 | 8 |
| 3 | 13 | 39 |
| 4 | 54 | 216 |
| 5 | 96 | 480 |
| 6 | 86 | 516 |
| 7 | 41 | 287 |
| 8 | 33 | 264 |
| 9 and over | 0 | 0 |
| | 328 | 1811 |

It should be noted that the first column is the class, or independent variable; the second column is the frequency, or dependent variable; and the third column is obtained by multiplying column (1) by column (2). The total of column (3) is the total number of rooms which, when divided by the total of column (2) (the number of dwellings), gives the mean number of rooms.

$$\frac{1811}{328} = 5.5 \text{ rooms}$$

EXAMPLE 2.3. GROUPED DATA. The deaths recorded according to age at death, in an investigation into the mortality of persons insured under annuity contracts were:

| Age Group | Number of Deaths |
|---|---|
| 50–54 | 16 |
| 55–59 | 58 |
| 60–64 | 180 |
| 65–69 | 513 |
| 70–74 | 1075 |
| 75–79 | 1748 |
| 80–84 | 1975 |
| 85–89 | 1569 |
| 90–94 | 600 |
| 95–99 | 183 |

Ages are recorded as age last birthday. Calculate the mean age at death.

*Solution.* Age group 50–54 includes all ages from the 50th birthday to the day before the 55th birthday, and hence, the exact age for the midpoint of this class interval is $52\frac{1}{2}$. The midpoint of successive classes is $57\frac{1}{2}$, $62\frac{1}{2}$, $67\frac{1}{2}$, etc. The assumption is made that each death occurs at the age corresponding to the midpoint of the class.

The calculation of the mean then proceeds as follows:

| Age Group (1) | Mid-Age Class (2) | Number of Deaths (3) | Col. (2) × Col. (3) (4) |
|---|---|---|---|
| 50–54 | $52\frac{1}{2}$ | 16 | 840 |
| 55–59 | $57\frac{1}{2}$ | 58 | 3,335 |
| 60–64 | $62\frac{1}{2}$ | 180 | 11,250 |
| 65–69 | $67\frac{1}{2}$ | 513 | $34{,}627\frac{1}{2}$ |
| 70–74 | $72\frac{1}{2}$ | 1075 | $77{,}937\frac{1}{2}$ |
| 75–79 | $77\frac{1}{2}$ | 1748 | 135,470 |
| 80–84 | $82\frac{1}{2}$ | 1975 | $162{,}937\frac{1}{2}$ |
| 85–89 | $87\frac{1}{2}$ | 1569 | $137{,}287\frac{1}{2}$ |
| 90–94 | $92\frac{1}{2}$ | 600 | 55,500 |
| 95–99 | $97\frac{1}{2}$ | 183 | $17{,}842\frac{1}{2}$ |
| | | 7917 | $637{,}027\frac{1}{2}$ |

$$\text{Mean age at death} = \frac{637{,}027\tfrac{1}{2}}{7{,}917} = 80.5 \text{ years}$$

**2.3. Use of Arbitrary Starting Point and Change of Scale.** Without a calculating machine the work of calculating the mean age at death in Example 2.3 would be laborious. Two changes in procedure simplify the calculation considerably. The first is the use of an *arbitrary starting point*. Instead of using age zero as the starting point of our calculations, we could call the midpoint of the first class zero. The midpoint of the second class would then be "age" 5 (i.e., $57\frac{1}{2} - 52\frac{1}{2}$), the next "age" 10, etc. By this means we calculate the mean "age" at death using $52\frac{1}{2}$ as age zero. When the calculations are complete, it must be remembered that this assumption has been made and that the mean "age," as calculated, must be increased by $52\frac{1}{2}$ to obtain the true age at death. However, it will be found that it simplifies the arithmetic still further if the calculations are made using as "age" zero the mid-age of the class which is near the middle of the range of values. Thus, if $77\frac{1}{2}$ is selected as "age" zero, the "age" to be used for the first group will be $52\frac{1}{2} - 77\frac{1}{2}$ or $-25$ and the complete range of "ages" will be $-25, -20, -15, -10, -5, 0, 5, 10, 15, 20$.

This compares with

$$0,\ 5,\ 10,\ 15,\ 20,\ 25,\ 30,\ 35,\ 40,\ 45$$

A further simplification can be obtained by *changing the scale*. It will be noted that the "ages" are all multiples of 5. If we use a new scale 5 times larger, the ages in the new units will be

$$-5,\ -4,\ -3,\ -2,\ -1,\ 0,\ 1,\ 2,\ 3,\ 4$$

The arithmetic is now reduced to manageable proportions as will be seen in Example 2.4. It is easy to show mathematically that these procedures do not alter the final result in any way.

EXAMPLE 2.4. Using an arbitrary starting point and a change in scale, calculate the mean age at death for the data given in Example 2.3.

*Solution.* The starting point will be taken as the mid-age of the group 75–79, i.e., age $77\frac{1}{2}$ and the scale of 5 years will be used.

The calculation with these units proceeds as follows:

| Age (1) | Midpoint of New Class Unit (2) | Number of Deaths (3) | Col. (2) × Col. (3) (4) |
|---|---|---|---|
| 50–54 | −5 | 16 | −80 |
| 55–59 | −4 | 58 | −232 |
| 60–64 | −3 | 180 | −540 |
| 65–69 | −2 | 513 | −1026 |
| 70–74 | −1 | 1075 | −1075 |
| 75–79 | 0 | 1748 | 0 |
| 80–84 | 1 | 1975 | 1975 |
| 85–89 | 2 | 1569 | 3138 |
| 90–94 | 3 | 600 | 1800 |
| 95–99 | 4 | 183 | 732 |
| | | 7917 | −2953 |
| | | | +7645 |
| | | *Total* | +4692 |

$$\text{Mean (in new units)} = \frac{4692}{7917} = .59$$

The mean age is then the starting point, which was 77½, plus the mean in the new units (.59) adjusted to the original scale, that is multiplied by 5, since the unit is 5 years.

$$\begin{aligned}\text{Mean age at death} &= 77.5 + (.59 \times 5) \\ &= 77.5 + 3.0 \\ &= 80.5 \text{ years}\end{aligned}$$

EXAMPLE 2.5. The following data are available for the incomes of Pennsylvania residents for 1960.

| Adjusted Gross Income Classes | Number of Returns | Total Adjusted Gross Income |
|---|---|---|
| Under $1,000 | 79,120 | $ 66,374,000 |
| $1,000–$1,999 | 292,734 | 440,117,000 |
| $2,000–$2,999 | 376,775 | 953,370,000 |
| $3,000–$3,999 | 436,386 | 1,530,709,000 |
| $4,000–$4,999 | 505,013 | 2,279,477,000 |
| $5,000–$5,999 | 497,228 | 2,731,442,000 |
| etc. | etc. | etc. |

Calculate the mean income of those residents whose adjusted gross income was less than $5,000. When a portion of the data only is to be used, such data is said to be *truncated*.

*Solution.* In this example, the data provided give not only the number of returns in each class but also the total adjusted gross income for the class. The table gives an accurate figure for this latter item and this must be used rather than the value obtained by assuming that the midvalue of the class interval is appropriate for the class. Thus, if we had used the midvalue of the class \$2,000–\$2,999, that is, \$2,500, we should have obtained a total adjusted earned income for the class of

$$376{,}775 \times \$2{,}500, \text{ or } \$941{,}940{,}000$$

compared with the true value of

$$\$953{,}370{,}000.$$

While for most classes, the difference will be unimportant, the classes near the end of the table may show remarkably different results. Thus, for the first class, the total based on the midvalue is

$$79{,}120 \times \$500 = \$39{,}560{,}000$$

compared with the true value of

$$\$66{,}374{,}000.$$

Adding the number of returns and total adjusted income for all residents with gross incomes less than \$5,000, we have:

| **Adjusted Gross Income Class** | **Number of Returns** | **Total Adjusted Gross Income (000 omitted)** |
|---|---|---|
| Under \$1,000 | 79,120 | \$ 66,374 |
| \$1,000–\$1,999 | 292,734 | 440,117 |
| \$2,000–\$2,999 | 376,775 | 953,370 |
| \$3,000–\$3,999 | 436,386 | 1,530,709 |
| \$4,000–\$4,999 | 505,013 | 2,279,477 |
| Under \$5,000 | 1,690,028 | \$5,270,047 |

$$\text{Mean income} = \frac{\$5{,}270{,}047{,}000}{1{,}690{,}028} = \$3{,}118$$

EXAMPLE 2.6. UNEQUAL CLASS INTERVALS. The 1960 Housing Census of the United States gives the following data on gross rents of renter occupied homes:

| Gross Rent | Number of Houses (in thousands) |
|---|---|
| Less than $20 | 320 |
| $20– | 736 |
| $30– | 1,221 |
| $40– | 1,755 |
| $50– | 2,245 |
| $60 | 2,555 |
| $70– | 2,350 |
| $80– | 3,490 |
| $100– | 1,693 |
| $120– | 973 |
| $150– | 406 |
| $200 or more | 166 |

Calculate the mean value of the Gross Rent.

*Solution.* The midpoints of the class intervals are:

$10, $25, $35, $45, $55, $65, $75, $90, $110, $135, $175, $250

For the last class interval, judgment must be used in choosing the midpoint, but since the number of houses in the interval is small, the error resulting from a poor choice is unimportant. A mean of $250 is suggested. $65 is an appropriate starting point to assume and $5 intervals. The calculation now proceeds as follows:

| Gross Rent (1) | Midpoint of Class (2) | Midpoint with New Starting Point (3) | Midpoint in New Class Scale (4) | No. of Houses (in thousands) (5) | Col. (4) × Col. (5) (6) |
|---|---|---|---|---|---|
| Less than $20 | $ 10 | –$ 55 | –11 | 320 | –3,520 |
| $20– | 25 | –40 | – 8 | 736 | –5,888 |
| $30– | 35 | –30 | – 6 | 1,221 | –7,326 |
| $40– | 45 | –20 | – 4 | 1,755 | –7,020 |
| $50– | 55 | –10 | – 2 | 2,245 | –4,490 |
| $60– | 65 | 0 | 0 | 2,555 | 0 |
| $70– | 75 | 10 | 2 | 2,350 | 4,700 |
| $80– | 90 | 25 | 5 | 3,490 | 17,450 |
| $100– | 110 | 45 | 9 | 1,693 | 15,237 |
| $120– | 135 | 70 | 14 | 973 | 13,622 |
| $150– | 175 | 110 | 22 | 406 | 8,932 |
| $200 or more | 250 | 185 | 37 | 166 | 6,142 |
| | | | | 17,910 | –28,244 |
| | | | | | +66,083 |
| | | | | | 37,839 |

$$\text{Mean value} = \$65 + \left(\$5 \times \frac{37{,}839}{17{,}910}\right)$$

$$= \$65 + (\$5 \times 2.11) = \$76$$

**2.4. The Mean of Groups of Data.** It often happens that data are available in grouped form and the means of the individual groups are known. In order to find the mean of the total data combined, it is unnecessary to go back to the original detail since the mean of the combined data is equal to the *weighted average* of the individual data. The use of unweighted averages is incorrect and can produce many misleading results.

EXAMPLE 2.7. A study is made of the I.Q. (Intelligent Quotient) of students in a certain high school and the following results are obtained:

| Grade | Number of Students | Mean I.Q. |
|---|---|---|
| 9 | 150 | 90 |
| 10 | 130 | 97 |
| 11 | 120 | 95 |
| 12 | 100 | 110 |
| | 500 | |

What is the mean I.Q. of the whole school?

*Solution.*

| Grade (1) | Number of Students (2) | Mean I.Q. (3) | Mean I.Q. Weighted by No. of Students Col. (2) × Col. (3) (4) |
|---|---|---|---|
| 9 | 150 | 90 | 13,500 |
| 10 | 130 | 97 | 12,610 |
| 11 | 120 | 95 | 11,400 |
| 12 | 100 | 110 | 11,000 |
| | 500 | | 48,510 |

$$\text{Mean I.Q.} = \frac{48{,}510}{500} = 97$$

It should be noted that the mean of the mean I.Q. values for each of the four grades is

$$\frac{90 + 97 + 95 + 110}{4} = \frac{392}{4} = 98$$

which does not give a correct value for the mean I.Q. of the 500 students. The reason for this is that each grade has been given equal weight and not each student. While this "short cut" will often give a reasonably approximate answer, it will only give the correct answer if the number of students in each class is the same.

**2.5. Properties of the Mean.**

(*a*) The arithmetic mean is the most commonly used average.

(*b*) It can be readily calculated when data are on punched cards or tape and computed by hand fairly readily from grouped tabulated data.

(*c*) It may be treated algebraically (see below).

(*d*) The sum of the (signed) differences of the individual deviations from the mean is zero. (No other measure has this property.)

(*e*) The sum of the squares of the differences of the individual deviations from the mean is a minimum (e.g., it is less than the sum of the squares computed from any other point).

(*f*) The mean is greatly affected by extreme values.

**2.6. Algebraic Properties of the Mean.**

If individual values are

$$x_1, x_2, x_3, \ldots, x_i, \ldots, x_n$$

then the mean is

$$x = \frac{1}{n}(x_1 + x_2 + x_3 + \cdots + x_i + \cdots + x_n)$$

The expression in brackets is represented mathematically by

$$\sum_{i=1}^{n} x_i$$

$\Sigma$ is a capital Greek *S*, or Sigma, as the letter is called, and indicates summation; $i = 1$ below the sigma gives the item being summed ($i$) and the first term of the summation (1). The $n$ above the sigma gives the final term.

When the item being summed and the range of terms is clear, the expression is often written simply

$$\Sigma x$$

Hence

$$(1) \quad \bar{x} = \frac{1}{n}\sum_{i=1}^{n} x_i$$

$$(2) \quad \bar{x} = \frac{1}{n}\sum_{i=1}^{n} (x_i - A) + A$$

for any $A$, which justifies the use of an arbitrary starting point, and

$$(3) \quad \bar{x} = \frac{1}{kn}\sum_{i=1}^{n} (kx_i)$$

for any $k$, which justifies the use of a change of scale. Writing $\bar{x}$ for $A$ in formula (2) gives

$$\frac{1}{n}\sum_{i=1}^{n} (x_i - \bar{x}) = 0$$

proving property (*d*).

If $\overline{X}_1$ is the mean of

$$x_1, x_2, x_3, \ldots, x_n \quad (n \text{ terms})$$

and $\overline{X}_2$ is the mean of

$$x_{n+1}, x_{n+2}, \ldots, x_{n+m} \quad (m \text{ terms})$$

then

$$n\overline{X}_1 = \sum_{i=1}^{n} x_i \text{ and } m\overline{X}_2 = \sum_{i=n+1}^{n+m} x_i$$

and hence

$$\frac{n\overline{X}_1 + m\overline{X}_2}{n + m} = \frac{1}{n + m}\sum_{i=1}^{n+m} x_i$$

proving that the mean of grouped data is the weighted average of the means of the individual groups.

Also, for *two groups the same size*, the sum of the means equals the mean of the sums.

For grouped data

$$\text{Mean} = A + \frac{\Sigma f_i d_i}{n} \quad \text{or } A + C\frac{\Sigma f_i u_i}{n}$$

where

$A$ = the arbitrary origin
$C$ = the class interval

$n$ = the total number of units
$f_i$ = the frequency of class $i$
$d_i$ = the deviation of the midpoint of class $i$ from the arbitrary origin $A$
$u_i$ = the deviation of the midpoint of class $i$, *expressed in class intervals*, from the arbitrary origin $A$.

## Problems

**Problem 2.1.** The following was the standing of the clubs in the National Hockey League on a certain date:

| Club | Won | Lost | Tied | Points |
|---|---|---|---|---|
| Chicago | 38 | 15 | 12 | 88 |
| Toronto | 29 | 25 | 11 | 69 |
| Rangers | 28 | 26 | 12 | 68 |
| Montreal | 27 | 25 | 13 | 67 |
| Detroit | 26 | 35 | 4 | 56 |
| Boston | 17 | 39 | 10 | 44 |

Calculate the mean number of games won, lost, and tied, and the mean number of points scored.

**Problem 2.2.** Explain why the mean number of games won equals the mean number of games lost in Problem 2.1.

**Problem 2.3.** Show that the mean number of points scored in Problem 2.1 can be obtained by the usual rule (2 for a win, 1 for a tie, and 0 for a loss) from the mean results and explain why this is so.

**Problem 2.4.** Calculate the mean age of the fifty-three students for which ages are given in Chapter 1, Example 1.2.

> *Note.* Remember to use an arbitrary starting point—age 20 is suggested. Also remember that the mean age as calculated will be the age last birthday, and hence, one-half must be added to this figure to obtain the true mean age.

**Problem 2.5.** Calculate the mean number of heads obtained by tossing six coins one hundred times in Problem 1.18. (Again remember to use an arbitrary starting point.)

**Problem 2.6.** What is the theoretical expected mean number of heads?

**Problem 2.7.** Calculate the mean inclination angle of lava flows in Hawaii from the data given in Example 1.1..

**Problem 2.8.** Baseball world series are won by the team which first wins four games. Hence, the number of games in a world series may be 4, 5, 6, or 7. The following table gives the distribution of the number of games in each world series from 1903 to 1961.

| **Number of games** | 4 | 5 | 6 | 7 | *Total* |
|---|---|---|---|---|---|
| **Number of series** | 10 | 14 | 15 | 19 = | 58 |

Calculate the mean number of games played in a series.

**Problem 2.9.** The following table gives an analysis of the number of reported accidents over the period 1956–1958, among a sample group of automobile drivers in California.

| **Number of Accidents** | **Number of Drivers** |
|---|---|
| 0 | 81,714 |
| 1 | 11,306 |
| 2 | 1,618 |
| 3 | 250 |
| 4 | 40 |
| 5 | 7 |
| | 94,935 |

Calculate the mean number of accidents per driver.

**Problem 2.10.** In Example 1.8, if $54\frac{1}{2}$ is taken as the arbitrary starting point and 10 as the scale unit, what is the midpoint of each class in the new class units.?

**Problem 2.11.** Calculate the mean grade per student in Example 1.8.

**Problem 2.12.** What would be an appropriate starting point and scale for the data in Problem 1.9?

**Problem 2.13.** Calculate the mean number of persons per room in owner occupied dwellings on the basis of the data in Problem 1.9.

**Problem 2.14.** Calculate the mean number of persons per room in renter occupied dwellings on the basis of the data in Problem 1.9.

**Problem 2.15.** The following table from a Department of Commerce study of automobile accidents published in 1964 gives the distribution of vehicle-miles by travel speed for day travel on main rural highways:

| MPH | Vehicle-miles (000,000 omitted) |
|---|---|
| 22 or less | 3 |
| 23–32 | 29 |
| 33–37 | 64 |
| 38–42 | 250 |
| 43–47 | 395 |
| 48–52 | 715 |
| 53–57 | 514 |
| 58–62 | 462 |
| 63–72 | 308 |
| 73 or more | 39 |

Calculate the mean travel speed.

**Problem 2.16.** A study of distribution by sex among litters of pigs gives the following distribution of male pigs in a litter of 5:

| Number of Male Pigs | Number of Litters |
|---|---|
| 0 | 2 |
| 1 | 20 |
| 2 | 41 |
| 3 | 35 |
| 4 | 14 |
| 5 | 4 |

Calculate the mean number of male pigs in five pig litters.

**Problem 2.17.** The 1960 housing census of the United States gives the following table of the value of vacant dwellings available for sale:

| Value (price asked) | No. of Dwellings (in thousands) |
|---|---|
| Less than $5,000 | 51 |
| $5,000–$9,900 | 97 |
| $10,000–$14,900 | 138 |
| $15,000–$19,900 | 104 |
| $20,000–$24,900 | 42 |
| $25,000 or more | 55 |

Calculate the mean value of these dwellings, assuming the mid-value of the first class interval is $2,450 and the mean value of the last class interval is $32,450.

**Problem 2.18.** The mean number of physician visits per person per year in the United States for the year July, 1963–June, 1964 is given in the following table. Using population weights given, calculate the mean number of physician visits for the whole population.

| Age | No. of Visits | Population (in millions) |
|---|---|---|
| Under 15 years | 3.8 | 56 |
| 15–24 years | 4.3 | 24 |
| 25–44 years | 4.5 | 47 |
| 45–64 years | 5.0 | 36 |
| 65–74 years | 6.3 | 11 |
| 75 + years | 7.3 | 6 |
| | | 180 |

**Problem 2.19.** The following table gives the number of accountants and their average earnings in 1964 from a study of salaries for selected occupations in private industry.

| Level of Responsibility | Number | Average Annual Salary |
|---|---|---|
| I | 4,000 | $ 6,250 |
| II | 8,500 | 7,000 |
| III | 18,000 | 8,000 |
| IV | 13,000 | 9,500 |
| V | 5,500 | 11,500 |

Calculate the average earnings of all accountants in the study.

**Problem 2.20.** In a study of accident involvement, data is subdivided by type of car, and day and night driving.

| | DAY | | NIGHT | |
|---|---|---|---|---|
| CAR | Vehicle-miles (millions) | Accident* Rate | Vehicle-miles (millions) | Accident* Rate |
| Small | 33 | 238 | 8 | 668 |
| Low-priced | 1,146 | 215 | 288 | 584 |
| Medium-priced | 813 | 196 | 190 | 591 |
| High-priced | 194 | 204 | 44 | 483 |
| | 2,186 | | 530 | |

*Per 100 million vehicle-miles.

Calculate the accident rate for each class of car for day and night driving combined.

## Solutions

**Problem 2.1.** $27\frac{1}{2}$ $27\frac{1}{2}$ $10\frac{1}{3}$ $65\frac{1}{3}$
**Problem 2.2.** The total number of games won must be equal to the total number of games lost, and since the mean number is the total number divided by the number of clubs, the mean number of games won must be equal to the mean number of games lost.
**Problem 2.3.** Giving 2 for a win, 1 for a tie, and 0 for a loss, we have

$$(27\frac{1}{2} \times 2) + (27\frac{1}{2} \times 0) + (10\frac{1}{3} \times 1) = 55 + 10\frac{1}{3} = 65\frac{1}{3}$$

the mean number of points scored.

Since this rule applies to the points for *each* individual team, the rule must apply to the sum of the points for all teams. Now the mean in each case is the sum divided by the number of teams and the rule must apply equally well to the means.
**Problem 2.4.** 20.7 years. Since ages are recorded at age last birthday, the class midpoint of students age 17 is 17.5.
**Problem 2.5.**

| No. of Heads (1) | Class Scale (2) | Number (3) | Col. (2) × Col. (4) |
|---|---|---|---|
| 0 | −3 | 2 | −6 |
| 1 | −2 | 7 | −14 |
| 2 | −1 | 26 | −26 |
| 3 | 0 | 29 | 0 |
| 4 | 1 | 24 | 24 |
| 5 | 2 | 10 | 20 |
| 6 | 3 | 2 | 6 |
| | | 100 | 50 |
| | | | −46 |
| | | | 4 |

$$\text{Mean} = 3 + \frac{4}{100} = 3.0 \text{ (approximately)}$$

**Problem 2.6.** Since heads and tails are equally likely, the theoretically expected number of heads on any throw is $\frac{1}{2}$. Since 6 throws are made at each trial, the expected number of heads is

$$\frac{1}{2} \times 6 = 3$$

**Problem 2.7.** 30°
**Problem 2.8.** 5.7 games
**Problem 2.9.** 0.16 accidents
**Problem 2.10.** $-4, -2, -1, 0, 1, 2, 3\frac{1}{2}$
**Problem 2.11.** Using the starting point and scale suggested in the previous problem:

| Mark (1) | New Class Midpoint (2) | Number of Students (3) | Col. (2) × Col. (3) (4) |
|---|---|---|---|
| Under 30 | −4 | 6 | −24 |
| 30–39 | −2 | 4 | −8 |
| 40–49 | −1 | 7 | −7 |
| 50–59 | 0 | 10 | 0 |
| 60–69 | 1 | 20 | 20 |
| 70–79 | 2 | 18 | 36 |
| 80–100 | 3½ | 5 | 17½ |
| | | 70 | 73½ |
| | | | −39 |
| | | | 34½ |

The mean number of marks is

$$54\frac{1}{2} + \left(10 \times \frac{34\frac{1}{2}}{70}\right) = 59\frac{1}{2}$$

**Problem 2.12.** Starting point = .88 and scale = .125. This gives class midpoints of

$$-5, \ -2, \ 0, \ 3, \ 8$$

(if we take the midpoint of the last group as 1.88).
**Problem 2.13.** Mean = 0.57
**Problem 2.14.** Mean = 0.71
**Problem 2.15.** The midpoint of the 6th group, 50, provides a suitable starting point and it is reasonable to assume 80 as the midpoint of the last group. Using 2½ as a unit, the class midpoints become:

$$-15, \ -9, -6, -4, -2, 0, 2, 4, 7, 12$$

(The first point is not exactly $-15$, but owing to the small volume of data in this group, this will not affect the results.)

$$\text{Mean speed} = 53 \text{ m.p.h.}$$

**Problem 2.16.** 2.4 pigs

**Problem 2.17.** $12{,}450 + \left(\frac{209}{487} \times 5000\right) = \$14{,}600$

**Problem 2.18.** 4.6 visits

**Problem 2.19.** $8,500. The calculated value is $8,474, but in view of the rounding of the original data, the answer should be rounded as indicated.

**Problem 2.20.** Small: 322. Low-priced: 289. Medium-priced: 271. High-priced: 256.

# 3

# Median, Mode, and Other Averages

**3.1. Median.** While the mean is the most commonly used average of a distribution and can be obtained readily with data processing equipment, it is not the only measure which can be used to indicate the central tendency of a body of data.

The *median* is defined as the middle value when data are arranged in an array according to size. If there are an even number of values, or if the data are grouped, it is necessary to interpolate to obtain the value of the median. These procedures are illustrated in Examples 3.2 and 3.3. The median divides the frequency histogram into two parts of equal area.

**3.2. Interpolation.** Interpolation is used extensively in statistical work. In its simplest form it can be very readily understood and applied. Suppose $y$ is a function of $x$, such that

| Value of x | Value of y |
|---|---|
| 20 | 13.7 |
| 30 | 15.9 |

It is required to find the value of $y$ corresponding to $x = 26$.

While $x$ increases from 20 to 30, i.e., by 10, $y$ increases from 13.7 to 15.9, i.e., by 2.2. It is reasonable to assume that when $x$ increases from 20 to 26, i.e., by 6, $y$ will increase by six-tenths of 2.2, i.e., by $(6/10) \times 2.2 = 1.3$. Hence, the value of $y$ corresponding to $x = 26$ is

$$13.7 + 1.3 = 15.0$$

The underlying assumption made is that the rate of increase of $y$ is constant over the range from one value of $x$ to the next, or in other words, the graph over this range is a straight line. While in certain circumstances this may not be a sufficiently accurate assumption, it is sufficient for most of the interpolations which are made in statistical work.

When the value required is halfway between the two values of $x$, the formula gives a figure halfway between the two values of $y$. Thus

$$\frac{5}{10} \times 2.2 = 1.1$$

and the mean is

$$13.7 + 1.1 = 14.8$$

This result can be obtained also by adding the two values and dividing by 2

$$\frac{13.7 + 15.9}{2} = 14.8$$

EXAMPLE 3.1. Calculate the median of the following range of numbers:

3, 4, 5, 7, 8, 9, 10, 12, 13, 15, 17

*Solution.* There are 11 values in all, and hence, the middle value is the sixth. The sixth value is 9 and this is the median value.

EXAMPLE 3.2. Calculate the median of the following range of numbers:

43, 57, 81, 92, 97, 105

*Solution.* There are 6 values in all, and hence there is no middle term. The middle of the range would lie halfway between the third term, 81, and the fourth term, 92. The function increases by $92 - 81 = 11$ over the range from 3 to 4, and hence will increase by half this amount, 11/2, over half the range. The median is

$$81 + \frac{11}{2} = 86\tfrac{1}{2}$$

Alternately, the median can be calculated as one half the sum of the two terms.

$$\frac{1}{2}(81 + 92) = 86\tfrac{1}{2}$$

**3.3 Median with Grouped Data.** With grouped data, the median will fall in one of the class intervals and although in many cases it may be sufficiently accurate to say that the midpoint of the class is the median, a more accurate procedure is generally used. The class in which the median falls is called the median class, and the assumption is made that individual values in the median class are distributed uniformly over the class. If

$L$ is the lower boundary of the median class
$U$ is the upper boundary of the median class
$f_i$ is the frequency of class $i$
$n$ is the total frequency
$m$ is the median class

then

$$\text{Median} = L + \frac{\frac{n}{2} - \sum_{0}^{m-1} f_i}{f_m}(U - L)$$

or expressed in words:

$$\text{Median} = \left[\begin{array}{l}\text{lower boundary} \\ \text{of median class}\end{array}\right] + \left[\begin{array}{l}\text{excess of half total} \\ \text{frequency over frequency} \\ \text{below median class}\end{array}\right] \times \left[\frac{\text{median class size}}{\text{median class frequency}}\right]$$

EXAMPLE 3.3. Calculate the median for the following data:

| Class | Number of Items (frequency) | Cumulative Frequency |
|---|---|---|
| 0– | 10 | 10 |
| 10– | 20 | 30 |
| 20– | 30 | 60 |
| 30– | 40 | 100 |
| 40– | 20 | 120 |
| 50– | 10 | 130 |
| | 130 | |

*Solution.* The median is halfway between the 65th and 66th value, and from the cumulative frequency table, lies in the fourth class (30–).

$$n = 130 \text{ and } \sum_{0}^{m-1} f_i = 60$$

$$\text{Median} = 30 + \left(\frac{65 - 60}{40} \times 10\right)$$

$$= 30 + \left(\frac{5}{40} \times 10\right) = 31\tfrac{1}{4}$$

**3.4. Mode.** Another useful measure of the center tendency is the mode. When there are a number of values at each point or in each class, the point or class with the greatest number is called the mode. When a frequency *curve* has been drawn, the mode is the maximum point on the graph. Occasionally there is no mode or more than one mode.

EXAMPLE 3.4. Find the mode of the following arrays of data.

(*a*) 2, 3, 3, 4, 4, 4, 5, 5, 5, 5, 6, 6, 7, 8
(*b*) 3, 5, 7, 9, 10, 11, 12, 13, 18
(*c*) 2, 3, 3, 3, 4, 4, 5, 6, 7, 7, 8, 8, 8, 9

*Solution.* (*a*) 5; (*b*) none; (*c*) 3 and 8.

**3.5 Mode for Grouped Data.** It is rarely necessary to use a more accurate figure than the midpoint of the modal class. More than one formula has been suggested for calculating a more accurate mode. The formula now most frequently used is

$$\text{Mode} = M + \frac{1}{2}\left(\frac{c - a}{2b - a - c}\right)C$$

where:

$M$ = the midpoint of the modal class
$C$ = class interval
$a$ = frequency of class immediately below modal class
$b$ = frequency of modal class
$c$ = frequency of class immediately above modal class

This can be expressed in the alternative form:

$$\text{Mode} = L + \left(\frac{b - a}{2b - a - c}\right)C$$

where $L$ = the lower boundary of modal class.

EXAMPLE 3.5. Using the data in Example 3.3, calculate the mode.
*Solution.* The mode lies in the fourth class and as a first approximation is the midpoint of that class, namely, 35. Calculated more accurately:

$$\text{Mode} = 35 + \frac{1}{2}\left(\frac{20 - 30}{80 - 20 - 30}\right) \times 10$$

$$\text{Mode} = 35 - \frac{1}{2}\left(\frac{10}{30}\right) \times 10$$

$$\text{Mode} = 33.3$$

**3.6. Relationship between Mean, Median, and Mode.** The median lies between the mean and the mode and is usually approximately one third of the way from the mean to the mode.

$$\text{Mean} - \text{Mode} = 3(\text{Mean} - \text{Median}) \tag{3.3}$$

With a symmetrical distribution, the mean, median, and mode will all be the same. If the distribution is skewed to the right, the mean will be to the right of the mode. Similarly, if the distri-

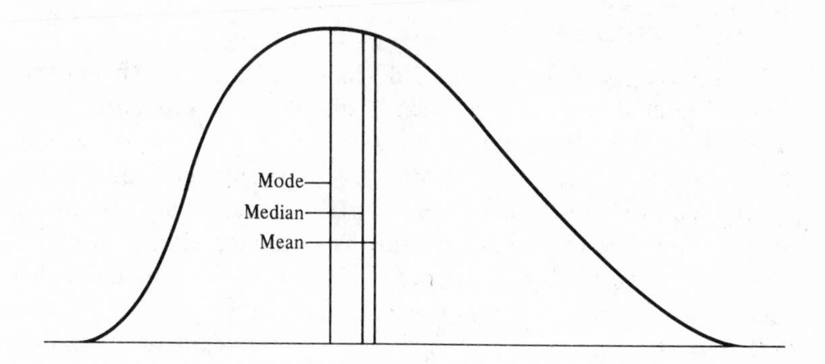


Figure 3.1. Distribution skewed to the right. Relative positions of mean, mode, and median.

bution is skewed to the left, the mean will be to the left of the mode. See Figure 3.1.

EXAMPLE 3.6. Calculate the mean for the distribution in Example 3.3 and show that the above rule is approximately correct in this case.

*Solution.* The mean is calculated as follows, using 35 as the starting point and 10 as the class interval.

| Class (1) | Midpoint of Class Scale (2) | Number of Items (3) | Col. (2) × Col. (3) (4) |
|---|---|---|---|
| 0– | −3 | 10 | −30 |
| 10– | −2 | 20 | −40 |
| 20– | −1 | 30 | −30 |
| 30– | 0 | 40 | 0 |
| 40– | 1 | 20 | 20 |
| 50– | 2 | 10 | 20 |
| | | 130 | −60 |

$$\text{Mean} = 35 - \left(10 \times \frac{60}{130}\right) = 35 - 4.6 = 30.4$$

$$\text{Median} = 31.3 \text{ (See Example 3.3)}$$

$$\text{Mode} = 33.3 \text{ (See Example 3.5)}$$

The distance between the mean and the mode is 2.9 and one-third of this is 1.0. The distance between the mean and median is 0.9.

**3.7. Advantages of the Median and Mode.** The *median*, being the point which divides the frequency distribution into two equal areas, has a readily understood significance as a measure of the central tendency. It is not distorted by unusual values and can be calculated even for "open ended" distributions, where all values in excess of a certain figure are thrown into one class. For this reason, it is used extensively by the United States Bureau of the Census for income, size of house, and similar distributions.

The *mode*, being the highest point on the frequency distribution, and hence, the most common value, has obvious significance. It is completely independent of extreme values and can be obtained readily for *small* distributions, if a mode exists. It becomes difficult to calculate for large volumes of data.

Neither the median nor the mode lend themselves readily to mathematical handling, and the values for large groups cannot be obtained from the values for subgroups. They are used less frequently than the mean as measures of central tendency.

**3.8. Other Measures of Central Tendency.** The following means are comparatively rarely used in statistical work as measures of central tendency.

*Geometric Mean*

The geometric mean of $n$ numbers, $x_1, x_2, \ldots, x_n$ is the $n$th root of the numbers all multiplied together.

$$G = \sqrt[n]{x_1 x_2 x_3 \ldots x_n}$$

*Harmonic Mean*

The harmonic mean is the reciprocal of the mean of the reciprocals of the numbers.

$$\frac{1}{H} = \frac{1}{n} \sum_{i=1}^{n} \frac{1}{x_i}$$

*The Root Mean Square*

The root mean square (R.M.S.) or *quadratic mean* is the square root of the mean of the squares of the individual values.

$$(\text{R.M.S.})^2 = \frac{1}{n} \sum_{i=1}^{n} x_i^2$$

The Geometric Mean ($G$) is never less than the Harmonic Mean ($H$) and the Arithmetic Mean ($M$) never less than the Geometric Mean.

$$H \leq G \leq M$$

EXAMPLE 3.7. Calculate the Arithmetic, Geometric, and Harmonic Means, and the Root Mean Square of the two numbers 2 and 8.

*Solution.*

$$M = \frac{2 + 8}{2} = 5.0$$

$$G = \sqrt{2 \times 8} = \sqrt{16} = 4.0$$

$$\frac{1}{H} = \frac{1}{2}\left(\frac{1}{2} + \frac{1}{8}\right) = \frac{1}{2} \times \frac{5}{8} = \frac{5}{16}$$

$$H = \frac{16}{5} = 3.2$$

$$(\text{R.M.S.})^2 = \frac{1}{2}(2^2 + 8^2) = \frac{1}{2}(4 + 64)$$

$$\text{R.M.S.} = \sqrt{34} = 5.8$$

EXAMPLE 3.8. Calculate the various means specified in Example 3.7 for the five numbers.

$$1,\ 2,\ 3,\ 4,\ 5$$

*Solution.*

$$M = \frac{1}{5}(1 + 2 + 3 + 4 + 5) = \frac{15}{5} = 3.0$$

$$G = \sqrt[5]{1 \times 2 \times 3 \times 4 \times 5} = \sqrt[5]{120}$$

$$\log 120 = 2.07918$$

$$\log \sqrt[5]{120} = \frac{1}{5}\log 120 = 0.41584$$

$$G = \sqrt[5]{120} = 2.6 \text{ (approx.)}$$

$$\frac{1}{H} = \frac{1}{5}\left(\frac{1}{1} + \frac{1}{2} + \frac{1}{3} + \frac{1}{4} + \frac{1}{5}\right)$$

$$= \frac{1}{5}(1.0 + 0.5 + 0.33 + 0.25 + 0.2)$$

$$= \frac{1}{5}(2.28) = .456$$

$$H = \frac{1}{.456} = 2.2 \text{ (approx.)}$$

$$(\text{R.M.S.})^2 = \frac{1}{5}(1^2 + 2^2 + 3^2 + 4^2 + 5^2)$$

$$= \frac{1}{5}(1 + 4 + 9 + 16 + 25)$$

$$= \frac{1}{5}(55) = 11$$

$$\text{R.M.S.} = \sqrt{11} = 3.3$$

**3.9. Quartiles, Deciles, and Percentiles.** It will be recalled that the median divides the distribution into two equal parts so that all values on one side are less than the median and all values above are greater than the median.

In the same way, a distribution may be divided by three *quartiles* or nine *deciles* or by ninety-nine *percentiles.* Only a few percentiles are normally used. Thus 5th, 10th, 25th, 75th, 90th, and 95th percentiles might be recorded to give a very full indication of a distribution. The 50th percentile is, of course, the median and the 25th is the first (lower) quartile and the 75th is the third (upper) quartile.

EXAMPLE 3.9. The following table gives the distribution of male pensioners under a private pension plan according to age at nearest birthday on a certain date. Determine the median, quartiles, deciles, and the 5th and 95th percentile points.

| Age | Number | Age | Number | Age | Number |
|---|---|---|---|---|---|
| 60 | 2 | 70 | 5 | 80 | 5 |
| 61 | 1 | 71 | 6 | 81 | 3 |
| 62 | 0 | 72 | 9 | 82 | 3 |
| 63 | 6 | 73 | 9 | 83 | 1 |
| 64 | 5 | 74 | 7 | 84 | 1 |
| 65 | 4 | 75 | 7 | 85 | 4 |
| 66 | 10 | 76 | 4 | 86 | 1 |
| 67 | 13 | 77 | 8 | 87 | 1 |
| 68 | 7 | 78 | 6 | 88 | 2 |
| 69 | 8 | 79 | 7 | 89 | 1 |
| | | | | **Total** | 146 |

*Solution.* First calculate the percentage distribution and the cumulative percentage distribution.

| Age | % | Cumulative % | Age | % | Cumulative % | Age | % | Cumulative % |
|---|---|---|---|---|---|---|---|---|
| 60 | 1.4 | 1.4 | 70 | 3.4 | 41.7 | 80 | 3.4 | 88.3 |
| 61 | 0.7 | 2.1 | 71 | 4.1 | 45.8 | 81 | 2.1 | 90.4 |
| 62 | 0.0 | 2.1 | 72 | 6.2 | 52.0 | 82 | 2.1 | 92.5 |
| 63 | 4.1 | 6.2 | 73 | 6.2 | 58.2 | 83 | 0.7 | 93.2 |
| 64 | 3.4 | 9.6 | 74 | 4.8 | 63.0 | 84 | 0.7 | 93.9 |
| 65 | 2.7 | 12.3 | 75 | 4.8 | 67.8 | 85 | 2.7 | 96.6 |
| 66 | 6.8 | 19.1 | 76 | 2.7 | 70.5 | 86 | 0.7 | 97.3 |
| 67 | 8.9 | 28.0 | 77 | 5.5 | 76.0 | 87 | 0.7 | 98.0 |
| 68 | 4.8 | 32.8 | 78 | 4.1 | 80.1 | 88 | 1.4 | 99.4 |
| 69 | 5.5 | 38.3 | 79 | 4.8 | 84.9 | 89 | 0.7 | 100.1 |

The difference from 100.0% and cumulative total is due to rounding. The median corresponds to the cumulative 50% point, and is age 72. The quartiles correspond to the cumulative 25%, 50%, and 75% points and are ages 67, 72, and 77. The deciles are ages 65, 67, 68, 70, 72, 74, 76, 78, and 81. The 5th and 95th percentile points are ages 63 and 85.

EXAMPLE 3.10. Using the data in the previous example, determine the mean and the mode. What would be the value of the mode if the approximate relationship referred to in Section 3.6 held?

*Solution.* The mean age can be calculated either from the original data or the percentage distribution, using age 75 as the arbitrary starting point. The results will be the same.

| Unit | Number | Unit × Number | Unit | Number | Unit × Number |
|---|---|---|---|---|---|
| −15 | 2 | −30 | 0 | 7 | 0 |
| −14 | 1 | −14 | 1 | 4 | 4 |
| −13 | 0 | 0 | 2 | 8 | 16 |
| −12 | 6 | −72 | 3 | 6 | 18 |
| −11 | 5 | −55 | 4 | 7 | 28 |
| −10 | 4 | −40 | 5 | 5 | 25 |
| −9 | 10 | −90 | 6 | 3 | 18 |
| −8 | 13 | −104 | 7 | 3 | 21 |
| −7 | 7 | −49 | 8 | 1 | 8 |
| −6 | 8 | −48 | 9 | 1 | 9 |
| −5 | 5 | −25 | 10 | 4 | 40 |
| −4 | 6 | −24 | 11 | 1 | 11 |
| −3 | 9 | −27 | 12 | 1 | 12 |

| Unit | Number | Unit × Number | Unit | Number | Unit × Number |
|---|---|---|---|---|---|
| −2 | 9 | −18 | 13 | 2 | 26 |
| −1 | 7 | −7 | 14 | 1 | 14 |
| | | | **Totals** | 146 | 250 |
| | | | | | −603 |
| | | | | | −353 |

$$\text{Mean age} = 75 - \frac{353}{146} = 72.6$$

The data has a principal mode at age 67, a secondary mode at $72\frac{1}{2}$ and a number of minor modes.

With a mean of 72.6 and a median of 72 (more accurately, 71.7) the mode would be expected to be approximately:

$$\begin{aligned}\text{Mode} &= 72.6 - 3 \times (72.6 - 71.7)\\ &= 72.6 - 2.7\\ &= 69.9\end{aligned}$$

Owing to the multimodal nature of the distribution, this is not close to the principal mode at 67.

## Problems

**Problem 3.1.** Using the simple interpolation method described in Section 3.2, find the value corresponding to 2.4 of a variable for which the value corresponding to 2.0 is 108 and the value corresponding to 3.0 is 123.

**Problem 3.2.** If the value of $y$ corresponding to $x = 10$ is 27 and the value of $y$ corresponding to $x = 20$ is 38, what is the value of $x$ corresponding to $y = 30$?

**Problem 3.3.** Given the following table of data, calculate the value of $y$ corresponding to $x = 2.5$ from (1) the values of $x = 1$ and 4; (2) the values of $x = 2$ and 3; (3) the values of $x = 1$ and 3 and (4) the values of $x = 2$ and 4.

| x | y |
|---|---|
| 1 | 256 |
| 2 | 324 |
| 3 | 400 |
| 4 | 484 |

**Problem 3.4.** Explain why the four answers to Problem 3.3 are not the same. Which estimate is the most accurate?
**Problem 3.5.** Calculate the median of the data in Example 1.1.
**Problem 3.6.** Calculate the median of the data in Example 1.4.
**Problem 3.7.** Calculate the median of the data in Example 1.8.
**Problem 3.8.** What is the mode of the data in Example 1.2?
**Problem 3.9.** What is the mode of the data in Example 1.1?
**Problem 3.10.** What is the median and the mode of the data in Example 2.2?
**Problem 3.11.** The following data are taken from a sample study made in 1963–4 of total short-stay hospital patients distributed according to age. Calculate the median age and the mode age of patients from these data.

| Age | Number of Patients (in thousands) |
|---|---|
| Under 15 | 4021 |
| 15–24 | 4083 |
| 25–44 | 7081 |
| 45–64 | 5806 |
| 65–74 | 2299 |
| 75 and over | 1547 |
| | 24837 |

**Problem 3.12.** What is the mean, median, and mode of the following theoretical results of tossing six coins sixty-four times?

| Number of Heads | Number of Occurrences |
|---|---|
| 0 | 1 |
| 1 | 6 |
| 2 | 15 |
| 3 | 20 |
| 4 | 15 |
| 5 | 6 |
| 6 | 1 |
| | 64 |

**Problem 3.13.** Find the median and mode of the speeds given in Problem 2.15.
**Problem 3.14.** Find the median and mode of the number of male pigs per litter in Problem 2.16.

**Problem 3.15.** Find the median and mode of the price of dwellings from the data in Problem 2.17.

**Problem 3.16.** Calculate the Arithmetic, Geometric, and Harmonic Means and the Root Mean Square for the following data.

$$1,\ 1,\ 2,\ 3,\ 4$$

**Problem 3.17.** Calculate the Harmonic Mean of

$$\frac{1}{2} \text{ and } \frac{1}{4}$$

Express the answer as a fraction.

**Problem 3.18.** Calculate the Geometric Mean of

$$2^2 = 4, \text{ and } 2^4 = 16$$

What is the answer expressed as a power of 2?

**Problem 3.19.** What are the quartiles of the data in Problem 3.12?

**Problem 3.20.** Given that the deciles of a certain distribution are as follows:

| **0%** | **10%** | **20%** | **30%** | **40%** | **50%** | **60%** | **70%** | **80%** | **90%** | **100%** |
|---|---|---|---|---|---|---|---|---|---|---|
| 0 | 2 | 6 | 14 | 26 | 46 | 70 | 90 | 97 | 99 | 100 |

Draw a frequency curve of the distribution.

## Solutions

**Problem 3.1.** $108 + \left(\frac{4}{10} \times 15\right) = 114$

**Problem 3.2.** $10 + \left(\frac{3}{11} \times 10\right) = 12.7$

**Problem 3.3.** (1) 370; (2) 362; (3) 364; (4) 364.

**Problem 3.4.** If the points are plotted on a graph and marked $A, B, C, D$, then estimate (1) is obtained by drawing a straight line between $A$ and $D$, (2) between $B$ and $C$, (3) between $A$ and $C$, and (4) between $B$ and $D$.

Since $A, B, C, D$ are not in a straight line, these lines will not all cut $x = 2.5$ at the same point. Estimate (2) is the most accurate because it is based on the two values *nearest* to $x = 2.5$.

**Problem 3.5.** $38\frac{1}{2}$

**Problem 3.6.** $69\frac{1}{2} + \dfrac{(32\frac{1}{2} - 31) \times 10}{21} = 70.2$

**Problem 3.7.** $59\frac{1}{2} + \dfrac{(35 - 27) \times 10}{20} = 63.5$

**Problem 3.8.** 20

**Problem 3.9.** With nearly every value different, the data really have no mode but technically it can be said that the data are bimodal with values at 30 and 47.

**Problem 3.10.**

$$\text{Median} = 4\tfrac{1}{2} + \frac{164 - 72}{96} = 4.5 + .96 = 5.5$$

$$\text{Mode} = 5$$

(With discrete data such as this, it is inappropriate to interpolate for the mode.)

**Problem 3.11.**

*Median.* The median class is age 25–44, and since these ages are presumably ages at last birthday (as normally recorded) the lower limit of the median class is age 25 exact (not $24\frac{1}{2}$) and the median age is

$$25 + \left(\frac{12418\frac{1}{2} - 8104}{7081}\right)20 = 37.2$$

*Mode.* It is tempting to assume that age 25–44 is the modal class since it has the largest number of patients. However, the class intervals are not equal and the number of patients per year of age in the second class is 408 and in the third class 354. Hence, the mode is probably in the range of ages 15 to 24 but it cannot be calculated accurately.

**Problem 3.12.** Since this data is symmetrical about the number 3, the mean, mode, and median will all be 3 and no calculations are necessary.

**Problem 3.13.**

$$\text{Median} = 47\tfrac{1}{2} + \left(\frac{1389\frac{1}{2} - 741}{715}\right) \times 5 = 52.0$$

$$\text{Mode} = 50 + \frac{1}{2}\left(\frac{514 - 395}{1430 - 514 - 395}\right) \times 5 = 50.6$$

**Problem 3.14.**

$$\text{Median} = 1\tfrac{1}{2} + \frac{58 - 22}{41} = 2.4$$

Since a litter of pigs must contain an exact number, the mode is correctly expressed as 2 rather than 2.3 as developed by formula.

**Problem 3.15.**

$$\text{Median} = \$9{,}950 + \left(\frac{243\tfrac{1}{2} - 148}{138}\right) \times \$5{,}000 = \$13{,}400$$

$$\text{Mode} = \$12{,}450 + \frac{1}{2}\left(\frac{104 - 97}{276 - 104 - 97}\right) \times \$5{,}000 = \$12{,}700$$

**Problem 3.16.**

*Arithmetic Mean*

$$M = \frac{1}{5}(1 + 1 + 2 + 3 + 4) = \frac{11}{5} = 2.2$$

*Geometric Mean*

$$G = \sqrt[5]{1 \times 1 \times 2 \times 3 \times 4} = \sqrt[5]{24} = 1.9$$

$$\log 24 = 1.38021$$

$$\frac{1}{5}\log 24 = 0.27604$$

$$\text{antilog} = 1.89$$

*Harmonic Mean*

$$\frac{1}{H} = \frac{1}{5}\left(\frac{1}{1} + \frac{1}{1} + \frac{1}{2} + \frac{1}{3} + \frac{1}{4}\right)$$

$$\frac{1}{H} = \frac{12 + 12 + 6 + 4 + 3}{5 \times 12} = \frac{37}{60} = 0.617$$

$$H = 1.6$$

*Root Mean Square*

$$(\text{RMS})^2 = \frac{1}{5}(1^2 + 1^2 + 2^2 + 3^2 + 4^2) = \frac{31}{5} = 6.2$$

$$\text{RMS} = 2.5$$

**Problem 3.17.**

$$\frac{1}{H} = \frac{1}{2}(2 + 4) = \frac{6}{2} = \frac{3}{1}$$

$$H = \frac{1}{3}$$

**Problem 3.18.**

$$G = \sqrt{2^2 \times 2^4} = \sqrt{2^6} = 2^3 = 8$$

The geometric mean is the third power of 2.

**Problem 3.19.**

| Number of Heads | Number of Occurrences | Cumulative Number of Occurrences |
|---|---|---|
| 0 | 1 | 1 |
| 1 | 6 | 7 |
| 2 | 15 | 22 |
| 3 | 20 | 42 |
| 4 | 15 | 57 |
| 5 | 6 | 63 |
| 6 | 1 | 64 |

The quartiles correspond to the cumulative number of occurrences of $\frac{1}{4}$, $\frac{1}{2}$ and $\frac{3}{4}$ of 64, namely, 16, 32 and 48. Hence the

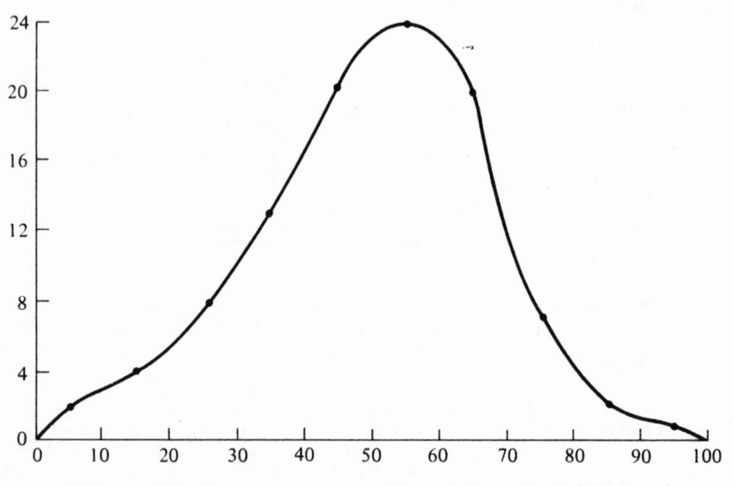


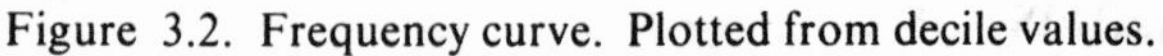
Figure 3.2. Frequency curve. Plotted from decile values.

quartiles are:

$$\text{First Quartile} = 1\tfrac{1}{2} + \frac{16 - 7}{15} = 2.1$$

$$\text{Second Quartile} = 2\tfrac{1}{2} + \frac{32 - 22}{20} = 3.0$$

$$\text{Third Quartile} = 3\tfrac{1}{2} + \frac{48 - 42}{15} = 3.9$$

**Problem 3.20.**

| Percentile Range | Midpoint | Number of Units |
|---|---|---|
| 0– | 5 | 2 |
| 10– | 15 | 4 |
| 20– | 25 | 8 |
| 30– | 35 | 12 |
| 40– | 45 | 20 |
| 50– | 55 | 24 |
| 60– | 65 | 20 |
| 70– | 75 | 7 |
| 80– | 85 | 2 |
| 90–100 | 95 | 1 |

See Figure 3.2.

# 4

# Dispersion—Standard Deviation

**4.1. Need for a Measure of Dispersion.** Two separate sets of data may contain the same number of items and have the same mean but one set may be much more dispersed or spread about the average value than the other. A measure of the *dispersion*, *scatter*, or *variation* from the mean is needed to help define the distribution more fully. The smaller the dispersion, the more typical the mean is of the whole distribution. See Figures 4.1 and 4.2.

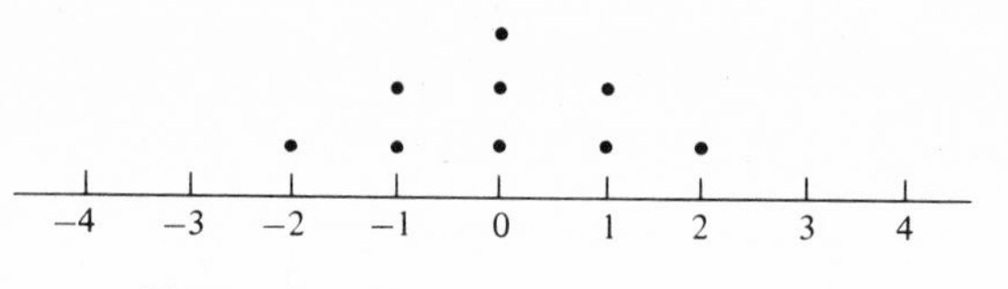


Figure 4.1. Dispersion. Small scatter.

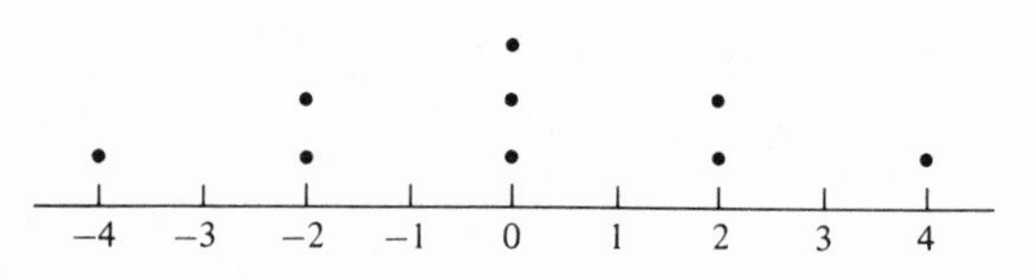


Figure 4.2. Dispersion. Large scatter.

**4.2. Measures of Dispersion.** A number of measures of dispersion are available.

1. The *Range*. The difference between the largest and the smallest values.
2. The *10–90 Percentile Range*. The difference between the 10th and the 90th percentile points.

Both these measures, as the word range implies, measure the

spread from one extreme to the other. The remaining measures of dispersion measure the departure from the mean or median, and hence measure one-half the spread.

3. The *Semi-interquartile Range* or *Quartile Deviation.* One-half the difference between the first and third quartiles.
4. The *Average Deviation from the Mean* or *Mean Deviation.* This is the arithmetic mean (average) of the individual absolute values of the deviations from the mean. See Section 4.6.
5. The *Standard Deviation* or its square, the *Variance.* This is the most generally used measure of dispersion and is discussed in the next section.
6. The *Half-width.* One-half of the width of the frequency curve at a height on the $y$ axis equal to one-half of the height of the modal point. (See Figure 4.3.)

In the illustration, $M$ is the modal point and $X$ is the midpoint of $MO$. $AB$ is the "width" of the curve and one-half $AB$ equals the "half-width."

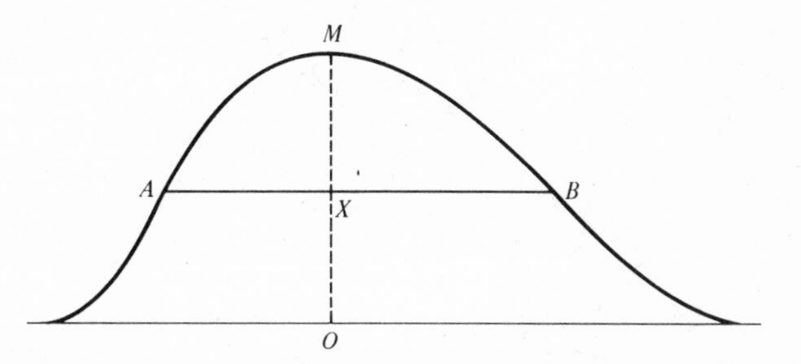


Figure 4.3. Frequency curve. Determination of half-width.

**4.3. Standard Deviation—Variance.** The *standard deviation* from the mean is the *quadratic mean* or *root mean square* of the deviations from the arithmetic mean. It will be recalled that the mean could be calculated by averaging the distances of the points from any origin. However, the standard deviation can be calculated, without adjustment, only from the *mean as origin.* The standard deviation is represented by the Greek letter $\sigma$ (sigma).* The *variance* ($\sigma^2$) is the square of the standard deviation.

*Some texts use $s$ instead of $\sigma$. See comments on the use of Latin and Greek letters in Chapter 5, Section 5.10.

With $n$ values, $x_1, x_2, x_3, \ldots, x_n$, the standard deviation is

$$\sigma = \sqrt{\frac{\sum_{i=1}^{n}(x_i - \bar{x})^2}{n}} \tag{4.1}$$

where $\bar{x}$ is the mean value of the $x$'s.

It will be noted that this formula involves (1) calculating the mean, (2) subtracting the mean from each individual value, (3) squaring each of the above results, (4) summing the squares, (5) dividing by the number of items involved, (6) taking the square root of this result. While a machine program can be devised for this procedure, a method of calculation is available which is easier for both machine and hand calculation.

The summation in Formula 4.1 can be written:

$$\begin{aligned}\sum_{i=1}^{n}(x_i - \bar{x})^2 &= \sum_{i=1}^{n}(x_i^2 - 2x_i\bar{x} + \bar{x}^2)\\ &= \sum_{i=1}^{n}x_i^2 - 2\bar{x}\sum_{i=1}^{n}x_i + n\bar{x}^2\\ &= \sum_{i=1}^{n}x_i^2 - n\bar{x}^2\end{aligned}$$

since

$$\frac{1}{n}\sum_{i=1}^{n}x_i = \bar{x}$$

by definition.

Hence, Formula 4.1 becomes

$$\sigma = \sqrt{\frac{\sum_{i=1}^{n}x_i^2}{n} - \bar{x}^2} \tag{4.2}$$

Since the standard deviation is obtained from a summation of the square of the distances of the individual values from the mean, the above formula is true whatever origin is used for determining the values of the $x$'s. Thus if the values measured from some arbitrary origin $A$ are $d_1, d_2, d_3 \ldots$, so that

$$d_i = x_i - A$$

then Formula 4.2 can be written

$$\sigma = \sqrt{\frac{\sum_{i=1}^{n} d_i^2}{n} - \bar{d}^2} \qquad (4.3)$$

where $\bar{d}$ will be the mean calculated from origin $A$ and

$$\bar{d} = \bar{x} - A$$

If $A$ is the mean $\bar{d}^2$ will be zero.

The use of a convenient arbitrary origin often simplifies the arithmetic involved in calculating the mean as will be apparent from the following example.

EXAMPLE 4.1. Calculate the mean and the standard deviation of the following five values.

3, 5, 6, 7, 10

*Solution.* Using Formula 4.1

| **Item** | **x** | **x − x̄** | **(x − x̄)²** |
|---|---|---|---|
| 1 | 3 | −3.2 | 10.24 |
| 2 | 5 | −1.2 | 1.44 |
| 3 | 6 | −0.2 | .04 |
| 4 | 7 | 0.8 | .64 |
| 5 | 10 | 3.8 | 14.44 |
| | 31 | | 26.80 |

$$\text{Mean} = \bar{x} = \frac{\Sigma x}{n} = \frac{31}{5} = 6.2$$

$$\text{Standard Deviation } (\sigma) = \sqrt{\frac{\Sigma (x - \bar{x})^2}{n}} = \sqrt{\frac{26.80}{5}}$$

$$= \sqrt{5.36} = 2.32$$

Using Formula 4.3, and taking 6 as the origin:

| Item (1) | Value x (2) | Distance from Assumed Origin d = (x − 6) (3) | Square of Column (3) $d^2$ (4) |
|---|---|---|---|
| 1 | 3 | −3 | 9 |
| 2 | 5 | −1 | 1 |
| 3 | 6 | 0 | 0 |
| 4 | 7 | 1 | 1 |
| 5 | 10 | 4 | 16 |
| *Divide by number of items.* | | 5 ) 1 | 5)27 |
| | | .2 | 5.4 |

$$\text{Mean} = 6 + .2 = 6.2$$

$$\text{Standard Deviation} = \sqrt{\Sigma\frac{d^2}{n} - \bar{d}^2}$$

$$= \sqrt{5.4 - (0.2)^2}$$

$$= \sqrt{5.36} = 2.32$$

It should be noted carefully that when using an arbitrary origin, the value of the origin has to be added to the average value calculated from that origin to obtain the mean, but no such adjustment is needed when using Formula 4.3 to calculate the standard deviation. In the above example, 6 is added to $\Sigma d/n =$ .2 to obtain the mean, but not to $\sqrt{5.36} = 2.32$ to obtain the standard deviation.

The square root of a number is frequently required in statistical work. This should be obtained from standard statistical tables.*

EXAMPLE 4.2. Calculate the mean and standard deviations of the following scores:

25, 17, 33, 18, 10, 15, 22, 20, 18, 21

---

*See: Arkin and Colton, *Tables for Statisticians*, 2nd ed., 1968, Barnes and Noble, College Outline Series.

*Solution.* Using 20 as a convenient origin

| Item | Score | Deviation from Origin | Square of Deviation |
|---|---|---|---|
| 1 | 25 | 5 | 25 |
| 2 | 17 | −3 | 9 |
| 3 | 33 | 13 | 169 |
| 4 | 18 | −2 | 4 |
| 5 | 10 | −10 | 100 |
| 6 | 15 | −5 | 25 |
| 7 | 22 | 2 | 4 |
| 8 | 20 | 0 | 0 |
| 9 | 18 | −2 | 4 |
| 10 | 21 | 1 | 1 |
| | | 21 | 10) 341 |
| | | −22 | 34.1 |
| | | 10 ) −1 | |
| | | −0.1 | |

$$\text{Mean} = 20 - 0.1 = 19.9$$

$$\text{Standard Deviation} = \sqrt{34.1 - (0.1)^2} = \sqrt{34.09} = 5.84$$

Note the great influence of the extreme values 33 and 10 on the standard deviation.

**4.4. Standard Deviation for Grouped Data.** For a large volume of data, Formula 4.3, with zero as origin, is most suitable for calculating the standard deviation when data processing equipment is available. However, data are often grouped in class intervals in order to assist the process of collection, to provide a presentation which can be more readily interpreted, or to assist in the hand calculation of statistical constants, such as the mean and the standard deviation. If the class interval is constant and equal to $C$, we may write $d_i = Cu_i$ where $u_i$ is the class unit variable having values 0, ±1, ±2, etc. Hence

$$x_i = A + Cu_i$$

where the values of $x$ are taken at the class midpoints, and the arbitrary origin $A$ must be a class midpoint.

Formula 4.3 becomes

$$\sigma = \sqrt{\frac{\Sigma f_i(Cu_i)^2}{n} - \left(\frac{\Sigma f_i Cu_i}{n}\right)^2}$$

$$\sigma = C\sqrt{\frac{\Sigma f_i u_i^2}{n} - \left(\frac{\Sigma f_i u_i}{n}\right)^2} \tag{4.4}$$

where $C$ = the class interval
$n$ = the total number of units
$f_i$ = the frequency of class $i$
$u_i$ = the deviation, in class intervals, of the midpoint of class $i$ from the arbitrary origin (which should be a midpoint of a class).

EXAMPLE 4.3. Calculate the standard deviation of the ages of students tabulated in Example 1.2.

*Solution.* Age 20 is a suitable origin (arbitrary starting point).

| Age | Number of Students (f) | Deviation from Origin (d) | Square of Deviation ($d^2$) | fd | $fd^2$ |
|---|---|---|---|---|---|
| 17 | 2 | −3 | 9 | −6 | 18 |
| 18 | 6 | −2 | 4 | −12 | 24 |
| 19 | 12 | −1 | 1 | −12 | 12 |
| 20 | 13 | 0 | 0 | 0 | 0 |
| 21 | 7 | 1 | 1 | 7 | 7 |
| 22 | 8 | 2 | 4 | 16 | 32 |
| 23 | 3 | 3 | 9 | 9 | 27 |
| 24 | 0 | 4 | 16 | 0 | 0 |
| 25 | 1 | 5 | 25 | 5 | 25 |
| 26 | 1 | 6 | 36 | 6 | 36 |
| | 53 | | | 53)13 | 53)181 |
| | | | | 0.2 | 3.42 |

$$\text{Standard Deviation} = \sqrt{3.42 - (0.2)^2}$$

$$= \sqrt{3.38} = 1.84$$

EXAMPLE 4.4. Calculate the standard deviation of the data of the deaths recorded in Example 2.3.

*Solution.* Using the midpoint (exact age $77\frac{1}{2}$) of the 6th group as origin and class intervals of 5 years:

| Age Group | Number of Deaths (f) | Midpoint of Class Unit (u) | $u^2$ | fu | $fu^2$ |
|---|---|---|---|---|---|
| 50–54 | 16 | −5 | 25 | −80 | 400 |
| 55–59 | 58 | −4 | 16 | −232 | 928 |
| 60–64 | 180 | −3 | 9 | −540 | 1620 |
| 65–69 | 513 | −2 | 4 | −1026 | 2052 |
| 70–74 | 1075 | −1 | 1 | −1075 | 1075 |
| 75–79 | 1748 | 0 | 0 | 0 | 0 |
| 80–84 | 1975 | 1 | 1 | 1975 | 1975 |
| 85–89 | 1569 | 2 | 4 | 3138 | 6276 |
| 90–94 | 600 | 3 | 9 | 1800 | 5400 |
| 95–99 | 183 | 4 | 16 | 732 | 2928 |
| | 7917 | | | 4692 | 22654 |
| | | *Divided by* 7917: | | .59 | 2.86 |

$$\text{Mean} = 77.5 + (5 \times .59) = 77.5 + 3.0 = 80.5 \text{ years}$$

$$\text{Standard Deviation} = 5\sqrt{2.86 - (.59)^2} = 5\sqrt{2.86 - .35}$$
$$= 5\sqrt{2.51} = 5 \times 1.58 = 7.9 \text{ years}$$

The 5 before the square root sign is the class interval.

**4.5. Characteristics of the Standard Deviation.** The standard deviation is the root square deviation about the mean. The standard deviation is less than the root square deviation about any point other than the mean. For the normal curve (see Chapter 7), 68.27% of the cases are included in the range from $\bar{x} - \sigma$ to $\bar{x} + \sigma$, that is to say, between the mean less the standard deviation and the mean plus the standard deviation. If the spread is increased to two standard deviations each side of the mean ($\bar{x} - 2\sigma$ to $\bar{x} + 2\sigma$), 95.45% of the cases will be included, and for three standard deviations each side of the mean, virtually all (99.73%) of the cases will be included. It must not be assumed that this is equally true of other distributions, but for many distributions these results will be approximately correct.

The standard deviation places great emphasis on extreme values because all individual deviations are squared in the calculations. Variances (the squares of the standard deviations) are additive. The variance of the sum of two independent variables is the sum of the variances of the two variables.

**4.6. Mean Deviation.** While the standard deviation is the most generally used measure of dispersion, other measures, enumerated in Section 4.2 are sometimes used. The mean deviation is the average of the deviation, regardless of sign, from the mean.

EXAMPLE 4.5. Calculate the mean deviation for the five values in Example 4.1.
*Solution:*

| Item | x | $x - \bar{x}$ | Deviation Regardless of Sign $\lvert x - \bar{x} \rvert$ |
|---|---|---|---|
| 1 | 3 | −3.2 | 3.2 |
| 2 | 5 | −1.2 | 1.2 |
| 3 | 6 | −0.2 | 0.2 |
| 4 | 7 | 0.8 | 0.8 |
| 5 | 10 | 3.8 | 3.8 |
| | | | 9.2 |

$$\text{Mean deviation} = \frac{9.2}{5} = 1.84$$

It will be noted that the mean deviation 1.84 is less than the standard deviation 2.32. For bell-shaped distributions, the mean deviation is about 80% of the standard deviation (79.79% in the case of the normal curve). While the mean, or average, deviation is normally calculated from the mean, it can be calculated from any other point. It is a minimum if calculated from the median.

**4.7. Relationship between Measures of Dispersion.** The following empirical relationships hold for many distributions:

(1) Mean deviation = 4/5 (standard deviation)
(2) Semi-interquartile range = 2/3 (standard deviation)

EXAMPLE 4.6. Find the relationship between the mean deviation and standard deviation of the data used for Examples 4.1 and 4.5.
*Solution.* The standard deviation was found to be 2.32 in Example 4.1, and the mean deviation was found to be 1.84 in

Example 4.5. The mean deviation was therefore

$$\text{mean deviation} = \frac{1.84}{2.32} \times \text{standard deviation}$$

$$\text{mean deviation} = .79 \times \text{standard deviation}$$

This is very close to the empirical figure of 0.8.

**4.8. Relative Measures of Dispersion.** The measures of dispersion described above are absolute values and are not, therefor, always suitable when comparisons have to be made between two distributions. Thus, a deviation of one ounce in measuring the weight of a man is unimportant; it is vital in measuring the dosage of a drug. A relative measure of dispersion is obtained by dividing the standard deviation by the mean and this is called the

$$\textit{Coefficient of Variation} = \frac{\sigma}{M} \text{ or } \frac{\sigma}{\bar{x}}$$

This measure is *dimensionless.*

EXAMPLE 4.7. What is the coefficient of variation of the five values in Example 4.1?
*Solution.* In Example 4.1 the mean was found to be 6.2, and the standard deviation was 2.32. Hence, the coefficient of variation = 2.32/6.2 = 0.37.

**4.9. Standardized Variables—Standard Scores.** When it is necessary to compare an individual value in one distribution with a corresponding individual value in another distribution we have to *standardize* the variables. Thus, if a student scores 80 in one test and 70 in another, it does not necessarily follow that he did better in the first test, relative to the other students in the class. The first test may have been easier. This difficulty of comparison is overcome by calculating a standardized variable

$$\frac{x - \bar{x}}{\sigma}$$

where $x$ is the individual value of the variable, $\bar{x}$ is the mean of the variable and $\sigma$ is the standard deviation. If the variable is a score in a test, the term *standard score* is often used for this expression.

To compare two distributions, it is useful to draw the relative frequency polynomials (See Chapter 1, Section 1.8) using standardized variables.

EXAMPLE 4.8. The following are the results of two tests:

| | Subject | |
|---|---|---|
| | **A** | **B** |
| Student $X$'s score | 80 | 70 |
| Mean score of class | 70 | 64 |
| Standard deviation of scores | 10 | 12 |

Calculate the standard scores for student $X$ in each subject.
*Solution.*

| | Subject | |
|---|---|---|
| | **A** | **B** |
| Standard score = | $\frac{80-70}{10}$ | $\frac{70-64}{12}$ |
| = | 1.0 | 0.5 |

The student had a better standard score in subject $A$. It will be noted that standard scores may be positive or negative. Scores above the mean are positive, those below, negative.

### 4.10. Corrections to Standard Deviation

1. *Small samples*

   When a small sample of a larger volume of data is being studied, the standard deviation is sometimes defined by writing $(n - 1)$ for $n$ in the denominator since this is a better estimate of the standard deviation of the population. However, this correction is usually ignored, and when $n$ is greater than 20 or 30, it is unimportant.

2. *Group data*

   The assumption that the data are located at the midpoint of each group is not strictly correct, since with a bell-shaped distribution, more of the data in each group will be nearer the higher point on the curve. Under certain theoretical conditions (where the curve goes gradually to zero in each direction) the correct value of the variance is obtained as follows:

$$\sigma^2 \text{ (correct value)} = \sigma^2 \text{ (from grouped data)} - \frac{C^2}{12}$$

where $C$ is the size of the class interval. This is called *Sheppard's Correction* for variance.

These two corrections are not generally used in practical work, and will not be used in the remainder of this book.

EXAMPLE 4.9. If the values in Example 4.1, namely 3, 5, 6, 7, 10, are a sample of larger data, what is the best estimate of the standard deviation of the population?

*Solution.* The value given in Example 4.1 was

$$\sqrt{\frac{26.80}{5}} = 2.32$$

The correct value is obtained by substituting $(n - 1) = 4$ for $n = 5$ in the denominator.

$$\sqrt{\frac{26.80}{4}} = 2.59$$

EXAMPLE 4.10. In Example 4.4, the standard deviation of age at death of a group of persons insured under annuity contracts was found to be $5\sqrt{2.51} = 7.9$ years. What is the standard deviation if Sheppard's Correction is applied?

*Solution.* The variance is the square of the standard deviation and is

$$25 \times 2.51$$

The class interval is 5 years.

Hence, the corrected value of the variance is

$$= (25 \times 2.51) - \frac{25}{12}$$

$$= 25\,(2.51 - .08)$$

$$= 25\,(2.43)$$

The standard deviation is

$$5\sqrt{2.43} = 5 \times 1.56 = 7.8 \text{ years}$$

It will be noted that the difference from 7.9 years calculated without the correction is very small.

**4.11. Charlier's Check.** If the standard deviation is calculated from two different arbitrary origins, the results should be the same; but if they differ, it is difficult to determine where the error lies. Charlier's Check uses two origins, one unit apart, and provides a check on the totals of the columns in the table rather than on the final result, thus simplifying the location of an error. The check is based on the relationship

$$f(u + 1)^2 = fu^2 + 2fu + f$$

Hence

$$\Sigma[f_i(u_i + 1)^2] = \Sigma[f_i u_i^2] + 2\Sigma(f_i u_i) + n$$

where $u_i$ = the deviation, in class intervals, of the midpoint of class $i$ from the arbitrary origin and $f_i$ is the frequency of class $i$. With the greater use of data processing equipment, the check is rarely used.

Example 4.11. Apply Charlier's Check to the calculation of the standard deviation in Example 4.4.

*Solution.*

| Age Group | Number of Deaths (f) | Midpoint of Class Unit (u) | (u + 1) | $(u + 1)^2$ | $f(u + 1)^2$ |
|---|---|---|---|---|---|
| 50–54 | 16 | −5 | −4 | 16 | 256 |
| 55–59 | 58 | −4 | −3 | 9 | 522 |
| 60–64 | 180 | −3 | −2 | 4 | 720 |
| 65–69 | 513 | −2 | −1 | 1 | 513 |
| 70–74 | 1075 | −1 | 0 | 0 | 0 |
| 75–79 | 1748 | 0 | 1 | 1 | 1748 |
| 80–84 | 1975 | 1 | 2 | 4 | 7900 |
| 85–89 | 1569 | 2 | 3 | 9 | 14121 |
| 90–94 | 600 | 3 | 4 | 16 | 9600 |
| 95–99 | 183 | 4 | 5 | 25 | 4575 |
| | 7,917 | | | | 39,955 |

Applying Charlier's Check

$$\begin{aligned} \Sigma fu^2 &= 22{,}654 \\ 2\Sigma fu = 2 \times 4{,}692 &= 9{,}384 \\ n &= 7{,}917 \end{aligned} \left.\vphantom{\begin{aligned}a\\a\\a\end{aligned}}\right\} \text{from previous calculations}$$

$$\text{Total} \quad 39{,}955$$

$$\Sigma f(u + 1)^2 = 39{,}955$$

which agree.

## Problems

**Problem 4.1.** The following are the speeds of 10 automobiles passing a certain check point:

40, 45, 45, 45, 50, 50, 50, 50, 55, 60

Find the range and the 10–90 percentile range from these speeds.
**Problem 4.2.** Find the semi-interquartile range for the speeds in Problem 4.1.
**Problem 4.3.** Find the average deviation from the mean of the speeds in Problem 4.1.
**Problem 4.4.** Find the standard deviation and the variance of the speeds in Problem 4.1.
**Problem 4.5.** The following were the scores of 100 students in a certain test:

19, 22, 27, 30, 32, 33, 35, 35, 36, 37
37, 38, 39, 40, 41, 42, 44, 45, 45, 45
46, 46, 47, 47, 48, 48, 49, 50, 50, 51
51, 51, 52, 52, 52, 53, 54, 54, 55, 56
57, 57, 58, 58, 59, 60, 60, 60, 61, 62
62, 62, 63, 63, 64, 64, 65, 65, 65, 65
65, 66, 66, 67, 68, 69, 70, 70, 71, 71
71, 72, 73, 74, 74, 75, 76, 76, 77, 77
77, 78, 78, 79, 79, 79, 80, 80, 81, 81
82, 83, 84, 86, 87, 87, 90, 90, 91, 93

Find the range, the 10–90 percentile range, and the semi-interquartile range of these data.
**Problem 4.6.** Find the standard and mean deviations of the data in Problem 4.5, grouping the data into ten mark classes so as to simplify the arithmetic.
**Problem 4.7.** Plot the data in Problem 4.5, and determine the half-width of the distribution.
**Problem 4.8.** Calculate the Standard Deviation of the data in Example 1.8. (The calculation of the mean will be found in the solution to Problem 2.11. This is an example of unequal intervals and the procedure for calculating the standard deviation under these circumstances is the same as for the mean.)
**Problem 4.9.** An analysis of the number of traffic violations among a group of drivers over a certain period of time was found to be as follows:

| Number of Traffic Violations | Number of Drivers |
|---|---|
| 0 | 27 |
| 1 | 25 |
| 2 | 12 |
| 3 | 5 |
| 4 | 1 |
| 5 and more | 0 |

Calculate the mean and the standard deviation of this distribution.

**Problem 4.10.** What is the mean deviation of the data in Problem 4.9?

**Problem 4.11.** The following is the distribution of the number of letters in the words in the paragraph from a novel:

| Number of Letters | Number of Words |
|---|---|
| 1 | 5 |
| 2 | 6 |
| 3 | 10 |
| 4 | 13 |
| 5 | 8 |
| 6 | 3 |
| 7 | 4 |
| 13 | 1 |

Find the mean, the standard deviation, and the mean deviation of the number of letters in a word.

**Problem 4.12.** It should be noted that with limited data such as that in the last question, the influence of the single extreme value may be quite large. The range of the data is 12, but if the thirteen letter word is omitted it would be only 6. Calculate the mean, the standard deviation, and the mean deviation of the data in Problem 4.11, omitting the thirteen letter word, and compare the results with the answers to Problem 4.11.

**Problem 4.13.** The employees of a certain firm receive the following weekly wages:

| Wages | Number of Employees |
|---|---|
| $ 55– | 10 |
| 65– | 12 |
| 75– | 15 |
| 85– | 20 |
| 95– | 14 |
| 105– | 7 |
| 115– | 2 |
| 125– | 0 |

Calculate the mean and standard deviation of the weekly wages.

**Problem 4.14.** What are the coefficients of variation of the data in Problems 4.1, 4.5, 4.9, 4.11, and 4.13?

**Problem 4.15.** If the independent variable assumes only positive values, what is the maximum and minimum possible values of the coefficient of variation for a symmetrical distribution?

**Problem 4.16.** Draw a relative frequency polygon of the data in Example 4.13, with standardized wages.

**Problem 4.17.** The results of three tests taken by a group of 10 students are set out below. Calculate the standard score of student $A$ in each test.

| | Test 1 | Test 2 | Test 3 |
|---|---|---|---|
| *A* | 70 | 70 | 70 |
| *B* | 70 | 65 | 75 |
| *C* | 65 | 80 | 50 |
| *D* | 40 | 50 | 60 |
| *E* | 65 | 75 | 70 |
| *F* | 75 | 80 | 80 |
| *G* | 65 | 65 | 65 |
| *H* | 60 | 70 | 60 |
| *I* | 50 | 50 | 40 |
| *J* | 60 | 65 | 60 |

**Problem 4.18.** For what volume of data in a small sample is the error, resulting from using $n$ instead of $n - 1$ in the denominator of the standard deviation, less than 5% of the standard deviation?

**Problem 4.19.** What is the effect of applying the Sheppard Correction to the standard deviations in the solutions to Problems 4.5 and 4.13?

**Problem 4.20.** Calculate the standard deviation of the following milk yield for cows:

| Milk Yield for Cow (gallons) | Number of Cows |
|---|---|
| 200– 299 | 10 |
| 300– 399 | 45 |
| 400– 499 | 90 |
| 500– 599 | 134 |
| 600– 699 | 124 |
| 700– 799 | 74 |
| 800– 899 | 59 |
| 900– 999 | 39 |
| 1000–1099 | 16 |
| 1100–1199 | 3 |

Check your calculations by Charlier's Check.

## Solutions

**Problem 4.1.** The maximum speed is 60 and the minimum is 40. The range is 60 – 40 or 20 m.p.h. The 10 percentile point lies halfway between 40 and 45 and can be taken to be $42\frac{1}{2}$. Similarly the 90 percentile point can be taken as $57\frac{1}{2}$. The 10–90 percentile range is $57\frac{1}{2} - 42\frac{1}{2} = 15$ m.p.h.

**Problem 4.2.** The quartiles are 45, 50, and 50, and the interquartile range is 50 – 45, or 5 m.p.h. The semi-interquartile range is $2\frac{1}{2}$ m.p.h.

**Problem 4.3.** To find the average deviation from the mean, the mean must first be calculated. Using 50 m.p.h. as the arbitrary starting point and class units of 5 m.p.h. gives:

| M.P.H. | u (1) | f (2) | fu Col. (1) × Col. (2) |
|---|---|---|---|
| 40 | −2 | 1 | −2 |
| 45 | −1 | 3 | −3 |
| 50 | 0 | 4 | 0 |
| 55 | 1 | 1 | 1 |
| 60 | 2 | 1 | 2 |
| | | 10 | −2 |

$$\text{Mean} = 50 + \left[5 \times \frac{(-2)}{10}\right]$$

$$\text{Mean} = 50 - 1 = 49 \text{ m.p.h.}$$

The average deviation can now be calculated.

| M.P.H. | Deviation from Mean (1) | Frequency (2) | Col. (1) × Col. (2) |
|---|---|---|---|
| 40 | 9 | 1 | 9 |
| 45 | 4 | 3 | 12 |
| 50 | 1 | 4 | 4 |
| 55 | 6 | 1 | 6 |
| 60 | 11 | 1 | 11 |
| | | 10 | 42 |

$$\text{Average deviation from the mean} = \frac{42}{10} = 4.2 \text{ m.p.h.}$$

If the calculation had been made from 50 m.p.h., the median, the answer would have been 4.0 m.p.h., verifying the statement in Example 4.5 that the average deviation from the median is less than the average deviation from the mean.

**Problem 4.4.** Using the same arbitrary starting point and class intervals as were used in the solution of the previous problem:

| M.P.H. | u (1) | f (2) | $u^2$ (3) | fu (1) × (2) | $fu^2$ (2) × (3) |
|---|---|---|---|---|---|
| 40 | −2 | 1 | 4 | −2 | 4 |
| 45 | −1 | 3 | 1 | −3 | 3 |
| 50 | 0 | 4 | 0 | 0 | 0 |
| 55 | 1 | 1 | 1 | 1 | 1 |
| 60 | 2 | 1 | 4 | 2 | 4 |
| | | 10 | | −2 | 12 |

$$\text{Mean} = 50 + 5\left(\frac{-2}{10}\right) = 49$$

$$\text{Standard Deviation} = 5\sqrt{\frac{12}{10} - \left(\frac{-2}{10}\right)^2} = 5\sqrt{1.2 - .04}$$

$$\text{Standard Deviation} = 5\sqrt{1.16} = 5 \times 1.08 = 5.4 \text{ m.p.h.}$$

The variance is the square of the standard deviation $= 25 \times 1.16 = 29$.

**Problem 4.5.** The range is $93 - 19 = 74$. The 10 and 90 percentile points are 37 and 81½ and hence the 10–90 percentile range is 44½. The quartile points are 48, 62, and 74½ and the semi-quartile range is:

$$\frac{74\frac{1}{2} - 48}{2} = 13\frac{1}{4}$$

**Problem 4.6.** Using $54\frac{1}{2}$ as the arbitrary starting point and 10 as the class interval:

| Score | u | f | $u^2$ | fu | $fu^2$ |
|---|---|---|---|---|---|
| 10–19 | −4 | 1 | 16 | −4 | 16 |
| 20–29 | −3 | 2 | 9 | −6 | 18 |
| 30–39 | −2 | 10 | 4 | −20 | 40 |
| 40–49 | −1 | 14 | 1 | −14 | 14 |
| 50–59 | 0 | 18 | 0 | 0 | 0 |
| 60–69 | 1 | 21 | 1 | 21 | 21 |
| 70–79 | 2 | 20 | 4 | 40 | 80 |
| 80–89 | 3 | 10 | 9 | 30 | 90 |
| 90–99 | 4 | 4 | 16 | 16 | 64 |
| | | 100 | | 63 | 343 |

$$\text{Mean} = 54\frac{1}{2} + (10 \times .63) = 54.5 + 6.3 = 60.8$$

$$\text{Standard deviation} = 10\sqrt{\frac{343}{100} - \left(\frac{63}{100}\right)^2}$$

$$= 10\sqrt{3.43 - .40}$$

$$= 10\sqrt{3.03} = 17.4$$

The mean deviation can be calculated without appreciable loss of accuracy from $60\frac{1}{2}$ rather than from 60.8, giving the distances for the mean for each class as $60\frac{1}{2} - 14\frac{1}{2}$, $60\frac{1}{2} - 24\frac{1}{2}$, etc.

| Score (1) | Difference from Mean (2) | f (3) | (2) × (3) |
|---|---|---|---|
| 10–19 | 46 | 1 | 46 |
| 20–29 | 36 | 2 | 72 |
| 30–39 | 26 | 10 | 260 |
| 40–49 | 16 | 14 | 224 |
| 50–59 | 6 | 18 | 108 |
| 60–69 | 4 | 21 | 84 |
| 70–79 | 14 | 20 | 280 |
| 80–89 | 24 | 10 | 240 |
| 90–99 | 34 | 4 | 136 |
| | | 100 | 1450 |

Mean deviation = 14.5

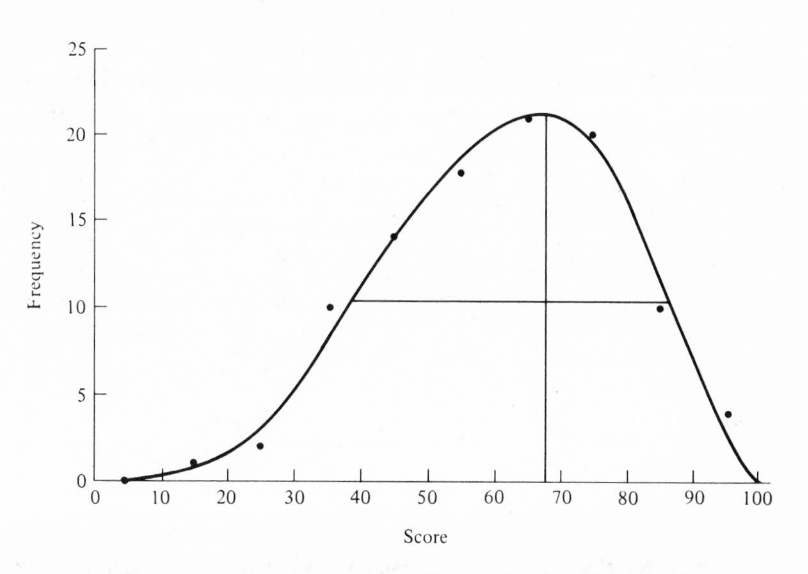


Figure 4.4. Frequency curve. Determination of half-width.

**Problem 4.7.** The graph is shown in Figure 4.4. Some judgment is needed to draw a smooth curve because the data is not very smooth. The width at one-half the height is seen to be about 48 and the half-width is 24.

**Problem 4.8.** Standard Deviation $= 10\sqrt{3.9 - (.5)^2} = 19$

**Problem 4.9.** 1 is the most convenient starting point for the calculation.

$$\text{Mean} = 1 - .03 = .97$$

$$\text{Standard Deviation} = \sqrt{.97 - (-.03)^2} = .98$$

**Problem 4.10.** Since the mean is very close to 1, the mean deviation can be calculated from 1 rather than from .97 without loss of accuracy. The mean deviation is .74.

**Problem 4.11.** Using 4 as the arbitrary starting point, the

$$\text{Mean} = 4 - \frac{2}{50} = 3.96 \text{ or } 4.0 \text{ approximately,}$$

and the

$$\text{Standard deviation} = \sqrt{4.32 - (.04)^2} = 2.1 \text{ approximately.}$$

The mean deviation is 1.4 approximately.

**Problem 4.12.** Using 4 as starting point, the mean = 3.8, the standard deviation = 1.6 and the mean deviation = 1.3. Note that the effect on the standard deviation is greater than on the mean deviation.

**Problem 4.13.** Using \$90 as an arbitrary starting point, and class intervals of \$10, the

$$\text{Mean} = \$90 + [10 \times (-.44)] = \$85.60$$

$$\text{Standard Deviation} = 10\sqrt{2.66 - (.44)^2} = 10\sqrt{2.66 - .19}$$

$$= 10\sqrt{2.47} = \$15.70$$

**Problem 4.14.** The coefficient of variation is the standard deviation divided by the mean $\left(\frac{\sigma}{M}\right)$.

| Problem Number | Mean | Standard Deviation | Coefficient of Variation |
|---|---|---|---|
| 4.1 | 49 | 5.4 | .11 |
| 4.5 | 60.8 | 17.4 | .29 |
| 4.9 | .97 | .98 | 1.01 |
| 4.11 | 4.0 | 2.1 | .53 |
| 4.13 | 85.60 | 15.70 | .18 |

**Problem 4.15.** If all values of the dependent variable correspond to a certain value ($x$) of the independent variable, then the mean is $x$, the standard deviation is zero, and the coefficient of variation is zero. If one-half of the values ($n/2$) of the dependent variable correspond to $x - t$, and the other half ($n/2$) to $x + t$, then the mean value is $x$ and the standard deviation is

$$\sqrt{t^2} = t$$

Now $t$ cannot be greater than $x$, otherwise $x - t$ would be negative. If $t = x$, the mean and the standard deviation of both $x$ and the coefficient of variation is 1. The coefficient of variation must lie in the range

$$0 \leq \text{coefficient of variation} \leq 1$$

**Problem 4.16.** The calculations required are as follows. From Problem 4.13, the mean is \$85.60 and the standard deviation is \$15.70.

| Wages | Class Midpoint (x) | x − x̄ | $\frac{x - \bar{x}}{\sigma}$ | Number of Employees | Relative Frequency |
|---|---|---|---|---|---|
| $ 55– | 60 | −25.60 | −1.63 | 10 | 12½% |
| 65– | 70 | −15.60 | −.99 | 12 | 15% |
| 75– | 80 | −5.60 | −.36 | 15 | 18¾% |
| 85– | 90 | 4.40 | .28 | 20 | 25% |
| 95– | 100 | 14.40 | .92 | 14 | 17½% |
| 105– | 110 | 24.40 | 1.55 | 7 | 8¾% |
| 115– | 120 | 34.40 | 2.19 | 2 | 2½% |
| | | | | 80 | 100% |

The frequency polygon is shown in Figure 4.5.

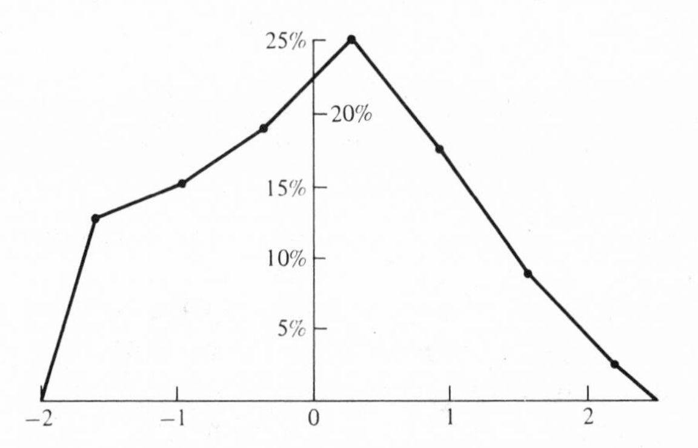


Figure 4.5. Relative frequency polygon with standardized variable.

**Problem 4.17.** Use 60 or 65 as the arbitrary starting point, and 5 as the class unit.

| | Test 1 | Test 2 | Test 3 |
|---|---|---|---|
| **Mean score** | 62 | 67 | 63 |
| **Standard Deviation** | 9.8 | 10.0 | 11.2 |
| **A's Score** | 70 | 70 | 70 |
| **A's Standard Score** | $\frac{8}{9.8}$ | $\frac{3}{10.0}$ | $\frac{7}{11.2}$ |
| | = 0.8 | = 0.3 | = 0.6 approx. |

**Problem 4.18.** The formula for the estimated standard deviation of the population is

$$\sigma = \sqrt{\frac{\Sigma(x - \bar{x})^2}{(n - 1)}}$$

If $n$ is used instead of $(n - 1)$ the error involved will be

$$\sqrt{\frac{\Sigma}{n-1}} - \sqrt{\frac{\Sigma}{n}}$$

If this is to be 5% of $\sqrt{\Sigma/(n - 1)}$

$$.05 \times \frac{1}{\sqrt{n-1}} = \frac{1}{\sqrt{n-1}} - \frac{1}{\sqrt{n}}$$

or

$$\frac{1}{\sqrt{n}} = \frac{.95}{\sqrt{n-1}}$$

Squaring both sides

$$(n - 1) = .9025n$$
$$.0975n = 1$$
$$n = 10.3$$

If the sample is over 10, the error is less than 5%.

**Problem 4.19.** The Sheppard Correction decreases the calculated variance by $C^2/12$ when $C$ is the class interval. The variance as calculated in Problem 4.5 was $100 \times 3.03 = 303$, and the corrected value is $303 - 100/12 = 303 - 8 = 295$. The corrected standard deviation is $\sqrt{295} = 17.2$.

The variance as calculated in Problem 4.13 was 247, and the corrected value is $247 - 100/12 = 239$. The correct standard deviation is $\sqrt{239}$ = \$15.50.

**Problem 4.20.** Using 649.5 as the arbitrary starting point, and class intervals of 100 gallons, the calculations are as follows:

| Yield | Class Midpoint (u) | Number of Cows (f) | $u^2$ | $(u+1)^2$ | fu | $fu^2$ | $f(u+1)^2$ |
|---|---|---|---|---|---|---|---|
| 200– 299 | −4 | 10 | 16 | 9 | −40 | 160 | 90 |
| 300– 399 | −3 | 45 | 9 | 4 | −135 | 405 | 180 |
| 400– 499 | −2 | 90 | 4 | 1 | −180 | 360 | 90 |
| 500– 599 | −1 | 134 | 1 | 0 | −134 | 134 | 0 |
| 600– 699 | 0 | 124 | 0 | 1 | 0 | 0 | 124 |
| 700– 799 | 1 | 74 | 1 | 4 | 74 | 74 | 296 |
| 800– 899 | 2 | 59 | 4 | 9 | 118 | 236 | 531 |
| 900– 999 | 3 | 39 | 9 | 16 | 117 | 351 | 624 |

| Yield | Class Midpoint (u) | Number of Cows (f) | $u^2$ | $(u+1)^2$ | fu | $fu^2$ | $f(u+1)^2$ |
|---|---|---|---|---|---|---|---|
| 1000–1099 | 4 | 16 | 16 | 25 | 64 | 256 | 400 |
| 1100–1199 | 5 | 3 | 25 | 36 | 15 | 75 | 108 |
| | | 594 | | | 388 | 2051 | 2443 |
| | | | | | −489 | | |
| | | | | | −101 | | |

$$\begin{aligned} \text{Check:}\quad \Sigma fu^2 &= 2051 \\ 2\Sigma fu &= -202 \\ n &= 594 \\ & \overline{\phantom{=}\ 2443} \\ \Sigma f(u+1)^2 &= 2443 \end{aligned}$$

$$\text{Mean Yield} = 649.5 - \left(100 \times \frac{101}{594}\right) = 632 \text{ gallons}$$

$$\text{Standard deviation} = 100\sqrt{\frac{2051}{594} - \left(\frac{101}{594}\right)^2}$$

$$= 100\sqrt{3.45 - (.17)^2}$$

$$= 100\sqrt{3.45 - .03} = 185 \text{ gallons}$$

# 5

# Moments, Skewness, and Kurtosis

**5.1. Moments.** In Chapter 2 it was shown that the most obvious single characteristic of a volume of data was the mean, and in Chapter 4 it was shown that the standard deviation provides a measure of the distribution of the data about the mean. It will be noted that the calculation of these two measures involves the summations

$$\frac{\Sigma fx}{n} \quad \text{and} \quad \frac{\Sigma fx^2}{n}$$

respectively, where $f$ is the frequency of the data for value $x$ and $n$ is the total volume of the data. These two summations are the first two terms of the series

$$\frac{\Sigma fx}{n}, \frac{\Sigma fx^2}{n}, \frac{\Sigma fx^3}{n}, \ldots, \frac{\Sigma fx^r}{n}, \ldots$$

These summations are called the first moment, the second moment, the third moment, . . . , the $r$th moment, . . . of the data. The moments are normally calculated about the origin but can be calculated about any point. $m_r'$ is the symbol for the $r$th moment about the origin and $m_r$ is the symbol for the $r$th moment about the mean. To be explicit, $m_r'$ is described as the "$r$th moment about zero," and $m_r$ is described as "the $r$th moment about the mean."

The expressions above are appropriate for grouped data, which is the usual form of data for which moments are calculated. When the data are not grouped, the $r$th moment becomes

$$m_r' = \frac{\sum_{i=0}^{n} x_i^r}{n}$$

EXAMPLE 5.1. Calculate the first four moments of the following distribution:

| x | f |
|---|---|
| 0 | 0 |
| 1 | 10 |
| 2 | 30 |
| 3 | 50 |
| 4 | 80 |
| 5 | 20 |
| 6 | 10 |
| 7 | 0 |
| | 200 |

*Solution.* The calculations can be made in tabular form as follows:

| x | x² | x³ | x⁴ | f | fx | fx² | fx³ | fx⁴ |
|---|---|---|---|---|---|---|---|---|
| 0 | 0 | 0 | 0 | 0 | 0 | 0 | 0 | 0 |
| 1 | 1 | 1 | 1 | 10 | 10 | 10 | 10 | 10 |
| 2 | 4 | 8 | 16 | 30 | 60 | 120 | 240 | 480 |
| 3 | 9 | 27 | 81 | 50 | 150 | 450 | 1350 | 4050 |
| 4 | 16 | 64 | 256 | 80 | 320 | 1280 | 5120 | 20480 |
| 5 | 25 | 125 | 625 | 20 | 100 | 500 | 2500 | 12500 |
| 6 | 36 | 216 | 1296 | 10 | 60 | 360 | 2160 | 12960 |
| 7 | 49 | 343 | 2401 | 0 | 0 | 0 | 0 | 0 |
| | | | | 200 | 700 | 2720 | 11380 | 50480 |
| Divide by $n$ = 200 | | | | | 3.5 | 13.6 | 56.9 | 252.4 |

The first four moments are, $m'_1 = 3.5$, $m'_2 = 13.6$, $m'_3 = 56.9$, and $m'_4 = 252.4$. Statistical tables* provide powers of $x$, but even so, the calculation of the higher moments is laborious. However, moments can be calculated very readily when data processing equipment is available.

**5.2. Moments about the Mean.** The mean is defined as

$$\bar{x} = \frac{\Sigma fx}{n}$$

*See Arkin and Colton, *Tables for Statisticians*, Barnes & Noble, Inc., New York, College Outline Series.

Hence

$$\frac{\Sigma f(x - \bar{x})}{n} = 0$$

and the first moment about the mean $m_1$ is zero. The variance is defined as

$$\sigma^2 = \frac{\Sigma f(x - \bar{x})^2}{n}$$

and hence, the second moment about the mean is equal to the variance

$$m_2 = \sigma^2$$

The standard deviation $\sigma$ equals $\sqrt{m_2}$.

The $r$th moment about the mean is

$$m_r = \frac{\Sigma f(x - \bar{x})^r}{n}$$

EXAMPLE 5.2. Calculate the first four moments about the mean of the data in Example 5.1.

*Solution.* From the results of Example 5.1, the mean of the distribution is 3.5, and the calculations can be set out in tabular form as follows:

| $x$ | $x - \bar{x}$ | $(x - \bar{x})^2$ | $(x - \bar{x})^3$ | $(x - \bar{x})^4$ | $f$ | $f(x - \bar{x})$ | $f(x - \bar{x})^2$ | $f(x - \bar{x})^3$ | $f(x - \bar{x})^4$ |
|---|---|---|---|---|---|---|---|---|---|
| 1 | −2.5 | 6.25 | −15.625 | 39.0625 | 10 | −25 | 62.5 | −156.25 | 390.6 |
| 2 | −1.5 | 2.25 | − 3.375 | 5.0625 | 30 | −45 | 67.5 | −101.25 | 151.9 |
| 3 | −0.5 | 0.25 | − 0.125 | .0625 | 50 | −25 | 12.5 | − 6.25 | 3.1 |
| 4 | 0.5 | 0.25 | 0.125 | .0625 | 80 | 40 | 20.0 | 10.00 | 5.0 |
| 5 | 1.5 | 2.25 | 3.375 | 5.0625 | 20 | 30 | 45.0 | 67.50 | 101.2 |
| 6 | 2.5 | 6.25 | 15.625 | 39.0625 | 10 | 25 | 62.5 | 156.25 | 390.6 |
| | | | | | 200 | 0 | 270.0 | − 30.00 | 1042.4 |
| | | Divide by $n = 200$ | | | | 0 | 1.35 | −0.15 | 5.21 |

The first four moments about the mean are, $m_1 = 0$, $m_2 = 1.35$, $m_3 = -0.15$ and $m_4 = 5.21$. It will be noted that the moments about the mean are much smaller than the moments about zero.

The calculation of moments about the mean is particularly laborious when the mean is not a round number, and even when the data processing equipment is used, two steps are required, since the mean must be calculated first. For these reasons, it is best to calculate the moments about an arbitrary origin, and then obtain the moments about the mean from these results by the use of the formulas in the following section.

**5.3. Relationship between Moments about Different Origins.** If $m_1$, $m_2$, $m_3$, etc., are the moments about the mean, and $m_1'$, $m_2'$, $m_3'$, etc., are moments about zero or *any other arbitrary origin*, then

$$\begin{aligned} m_1 &= 0 \\ m_2 &= m_2' - (m_1')^2 \\ m_3 &= m_3' - 3m_1'm_2' + 2(m_1')^3 \\ m_4 &= m_4' - 4m_1'm_3' + 6(m_1')^2m_2' - 3(m_1')^4 \end{aligned}$$

These formulas follow from the expansion of the expressions in Section 5.2. It will be noted that since $\sigma = \sqrt{m_2}$, the second equation can be written

$$\sigma = \sqrt{\frac{\sum_{i=0}^{n} x_i^2}{n} - \bar{x}^2}$$

which in Formula 4.2 in Chapter 4.

Example 5.3. Calculate the first four moments about the mean of the data in Example 5.1 by means of the formulas in Section 5.3.

*Solution.* In Example 5.1, the values of the moments about the origin were

$$m_1' = 3.5, m_2' = 13.6, m_3' = 56.9, m_4' = 252.4$$

Hence, by the formulas in Section 5.3

$$\begin{aligned} m_1 &= 0 \\ m_2 &= m_2' - (m_1')^2 = 13.6 - (3.5)^2 \\ &= 13.6 - 12.25 \\ &= 1.35 \end{aligned}$$

$$\begin{aligned} m_3 &= m_3' - 3m_1'm_2' + 2(m_1')^3 \\ &= 56.9 - (3 \times 3.5 \times 13.6) + 2(3.5)^3 \\ &= 56.9 - 142.8 + 85.75 \\ &= -.15 \end{aligned}$$

$$\begin{aligned} m_4 &= m'_4 - 4m'_1m'_3 + 6(m'_1)^2m'_2 - 3(m'_1)^4 \\ &= 252.4 - (4 \times 3.5 \times 56.9) + (6 \times (3.5)^2 \times 13.6) - 3(3.5)^4 \\ &= 252.4 - 796.6 + 999.6 - 450.2 \\ &= 5.2 \end{aligned}$$

These values are the same as those obtained in Example 5.2, where the values were calculated directly from the mean.

EXAMPLE 5.4. A coin is tossed until a head comes uppermost. The number of tosses may be 1 (head), 2 (tail, head), 3 (tail, tail, head), or more; 2048 throws are made and the results recorded as follows:

| Distribution of heads and tails | Number of tosses | Number of occasions |
|---|---|---|
| H | 1 | 1061 |
| TH | 2 | 494 |
| $T^2H$ | 3 | 232 |
| $T^3H$ | 4 | 137 |
| $T^4H$ | 5 | 56 |
| $T^5H$ | 6 | 29 |
| $T^6H$ | 7 | 25 |
| $T^7H$ | 8 | 8 |
| $T^8H$ | 9 | 6 |
| | | 2048 |

Calculate the first three moments of this distribution about the mean.

*Solution.* The moments about the arbitrary origin of 3 are first calculated:

| d | $d^2$ | $d^3$ | f | fd | $fd^2$ | $fd^3$ |
|---|---|---|---|---|---|---|
| −2 | 4 | −8 | 1061 | −2122 | 4244 | −8488 |
| −1 | 1 | −1 | 494 | −494 | 494 | −494 |
| 0 | 0 | 0 | 232 | 0 | 0 | 0 |
| 1 | 1 | 1 | 137 | 137 | 137 | 137 |
| 2 | 4 | 8 | 56 | 112 | 224 | 448 |
| 3 | 9 | 27 | 29 | 87 | 261 | 783 |
| 4 | 16 | 64 | 25 | 100 | 400 | 1600 |
| 5 | 25 | 125 | 8 | 40 | 200 | 1000 |
| 6 | 36 | 216 | 6 | 36 | 216 | 1296 |
| | | | 2048 | −2104 | 6176 | −3718 |

Moments about origin of 3 are

$$m_1' = -\frac{2104}{2048} = -1.027$$

$$m_2' = \frac{6176}{2048} = 3.02$$

$$m_3' = -\frac{3718}{2048} = -1.82$$

The mean is 3 − 1.027 = 1.973.

The moments about the mean, from the formula in Section 5.3, are

$$m_1 = 0$$

$$m_2 = 3.02 - (-1.027)^2$$
$$= 3.02 - 1.05 = 1.97$$

$$m_3 = -1.82 - 3(-1.027)(3.02) + 2(-1.027)^3$$
$$= -1.82 + 9.30 - 2.17 = 5.31$$

EXAMPLE 5.5. Check the results obtained in Example 5.4, by making the same calculation with zero as origin.

*Solution.*

| x | $x^2$ | $x^3$ | f | fx | $fx^2$ | $fx^3$ |
|---|---|---|---|---|---|---|
| 0 | 0 | 0 | 0 | 0 | 0 | 0 |
| 1 | 1 | 1 | 1061 | 1061 | 1061 | 1061 |
| 2 | 4 | 8 | 494 | 988 | 1976 | 3952 |
| 3 | 9 | 27 | 232 | 696 | 2088 | 6264 |
| 4 | 16 | 64 | 137 | 548 | 2192 | 8768 |
| 5 | 25 | 125 | 56 | 280 | 1400 | 7000 |
| 6 | 36 | 216 | 29 | 174 | 1044 | 6264 |
| 7 | 49 | 343 | 25 | 175 | 1225 | 8575 |
| 8 | 64 | 512 | 8 | 64 | 512 | 4096 |
| 9 | 81 | 729 | 6 | 54 | 486 | 4374 |
| | | | 2048 | 4040 | 11984 | 50354 |

Moments about origin of 0 are

$$m_1 = \frac{4040}{2048} = 1.973$$

$$m_2' = \frac{11984}{2048} = 5.85$$

$$m_3' = \frac{50354}{2048} = 24.59$$

Moments about the mean are

$$m_1 = 0$$

$$m_2 = 5.85 - (1.973)^2$$
$$= 5.85 - 3.89 = 1.96$$

$$m_3 = 24.59 - 3(1.973)(5.85) + 2(1.973)^3$$
$$= 24.59 - 34.63 + 15.36 = 5.32$$

**5.4. Moment Generating Function.** In theoretical work, considerable use is made of a function called the moment generating function. This function is of the form

$$M_x(\theta) = \sum_{x=0}^{\infty} e^{\theta x} f(x)$$

which when expanded gives

$$M_x(\theta) = 1 + \theta m_1' + \frac{\theta^2}{2!} m_2' + \frac{\theta^3}{3!} m_3' + \cdots$$

The $r$th moment about the origin is seen to be the coefficient of $\theta^r/r!$

**5.5. Skewness.** When a distribution is not symmetrical about its mean value, it is said to be *skew.* If the tail of the distribution is longer on the right of the mode, the distribution is said to be *skewed to the right*, or to have *positive skewness.* Similarly, if the tail is longer on the left, the distribution is *skewed to the left*, or has *negative skewness.* (See Figure 3.1).

**5.6. Measures of Skewness.** The three most common measures of skewness are

$$(1)\quad \frac{\text{mean} - \text{mode}}{\text{standard deviation}}$$

$$(2)\quad \frac{3(\text{mean} - \text{median})}{\text{standard deviation}}$$

$$(3)\ a_3 = \frac{m_3}{\sqrt{m_2^3}}$$

The first and second measures listed are sometimes referred to as *Pearson's first and second coefficients of skewness*, respectively. It will be noted that if the approximate relationship between the mean, mode, and median (Formula 3.3) is exact, (1) and (2) will be identical.

The third measure is called the *moment coefficient of skewness* ($a_3$), and since it is calculated from the moments about the mean, it can be readily obtained when data processing equipment is available. The symbol $b$ or $\beta$ (Greek beta), is sometimes used for $a_3^2$.

It will be noted that these measures are positive when the distribution is skewed to the right and negative when skewed to the left, justifying the use of the expressions negative and positive skewness.

EXAMPLE 5.6. Calculate the three measures of skewness listed above for the data in Example 5.1.

*Solution.* For these data the mean is 3.5 as calculated in Example 5.1. The data in this example takes on only discrete values 1, 2, 3, etc., but to calculate the mean and mode in order to determine the skewness, it should be assumed that the function is continuous and represents the range 1/2 to $1\frac{1}{2}$, etc.

Using the procedures for calculating the median and mode described in Chapter 3, Sections 3.3 and 3.5, the median is

$$3.5 + \left(\frac{100 - 90}{80}\right) \times 1 = 3.63$$

and the mode is

$$4 + \frac{1}{2}\left(\frac{20 - 50}{160 - 50 - 20}\right) \times 1 = 3.83$$

Hence, the three measures of skewness are

$$(1)\ \frac{3.5 - 3.83}{1.35} = \frac{-.33}{1.35} = -.24$$

$$(2)\ \frac{3(3.5 - 3.63)}{1.35} = \frac{3 \times (-.13)}{1.35} = -.29$$

$$(3)\quad \frac{-0.15}{\sqrt{(1.35)^3}} = \frac{-0.15}{\sqrt{2.46}} = \frac{-0.15}{1.57} = -.096$$

It will be noted that these measures are negative and the distribution is skewed to the left.

EXAMPLE 5.7. Calculate the moment coefficient of skewness of the data in Example 5.5.
*Solution.* For these data $m_2 = 1.96$ and $m_3 = 5.32$
The moment coefficient of skewness is

$$\frac{m_3}{\sqrt{m_2^3}} = \frac{5.32}{\sqrt{1.96^3}} = \frac{5.32}{\sqrt{7.53}} = \frac{5.32}{2.74} = 1.94$$

The distribution has a marked positive skewness.

**5.7. Kurtosis.** Kurtosis is the "peakedness" of a distribution. The *normal curve*, which has been referred to briefly and is described in Chapter 7, is taken as the standard of "peakedness." A curve less peaked than the normal is said to be *platykurtic* and a more peaked curve is said to be *leptokurtic* (see Figure 5.1). The term *mesokurtic* is sometimes used to describe the normal curve.

Like skewness, there is more than one measure of kurtosis. One measure of kurtosis is

$$\frac{Q}{P_{90} - P_{10}}$$

where $Q = 1/2\,(Q_3 - Q_1)$, the semi-quartile range, and $P_{90}$ and $P_{10}$ are the ninety and the ten percentile poihts.

The measure based on moments is the *moment coefficient of kurtosis* and makes use of the fourth moment. It is designated $a_4$, or $b_2$, or $\beta_2$ (Greek beta).

$$a_4 = \frac{m_4}{m_2^2}$$

For the normal curve, $a_4 = 3$, and hence if $a_4 < 3$ the curve is *platykurtic*, and if $a_4 > 3$ the curve is *leptokurtic*. (Sometimes $a_4 - 3$ is used as the measure of kurtosis.)

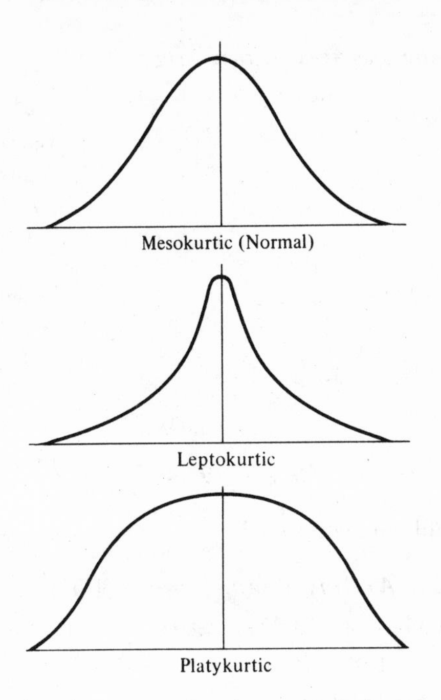


Figure 5.1. Frequency curves, showing normal, leptokurtic, and platykurtic distributions.

EXAMPLE 5.8. Calculate the moment coefficient of kurtosis of the data in Example 5.1. Is the distribution platykurtic or leptokurtic?

*Solution.* The second and further moments about the mean for these data, were calculated in the solution to Example 5.2. $m_2$ was 1.35 and $m_4$ was 5.21. The moment coefficient of kurtosis is

$$\frac{5.21}{(1.35)^2} = \frac{5.21}{1.82} = 2.9$$

Since this is less than 3, the distribution is very slightly platykurtic.

EXAMPLE 5.9. Calculate the moment coefficient of kurtosis of the data in Example 5.4.

*Solution.* Only the first three moments were calculated in the solution to Example 5.4 and hence, the fourth moment must be calculated. Using 3 as an arbitrary origin:

| d | $d^4$ | f | $fd^4$ |
|---|---|---|---|
| −2 | 16 | 1061 | 16976 |
| −1 | 1 | 494 | 494 |
| 0 | 0 | 232 | 0 |
| 1 | 1 | 137 | 137 |
| 2 | 16 | 56 | 896 |
| 3 | 81 | 29 | 2349 |
| 4 | 256 | 25 | 6400 |
| 5 | 625 | 8 | 5000 |
| 6 | 1296 | 6 | 7776 |
| | | 2048 | 40028 |

$$m'_4 = 19.54$$

From the formula in Section 5.3

$$\begin{aligned} m_4 &= m'_4 - 4\,m'_1 m'_3 + 6(m'_1)^2\,m'_2 - 3(m'_1)^4 \\ &= 19.54 - 4(-1.027)(-1.82) + 6(-1.027)^2(3.02) \\ &\quad - 3(-1.027)^4 \\ &= 19.54 - 4(1.87) + 6(3.185) - 3(1.112) \\ &= 19.54 - 7.48 + 19.11 - 3.34 \\ m_4 &= 27.83 \end{aligned}$$

Moment coefficient of kurtosis

$$= \frac{27.83}{(1.96)^2} = \frac{27.83}{3.84} = 7.2$$

The distribution is leptokurtic.

**5.8. Dimensionless Moments.** In Chapter 4, Section 4.8, the need for a measure of dispersion which did not depend upon absolute values was explained and the *coefficient of variation* ($\sigma/m$) was used for this purpose which is a *dimensionless* function.

When dimensionless moments are needed, the moments about the mean are divided by $(\sqrt{m_2})^r$ and are represented by $a_r$.

$$a_1 = 0$$

$$a_2 = 1$$

$$a_3 = \frac{m_3}{(\sqrt{m_2})^3} = \frac{m_3}{\sigma^3}$$

$$a_r = \frac{m_r}{(\sqrt{m_2})^r} = \frac{m_r}{\sigma^r}$$

It will be noted particularly that the dimensionless second moment is unity and *not the same* as the dimensionless coefficient of variation.

$a_3$, *the moment coefficient of skewness*, was introduced in Section 5.6, and $a_4$, the *moment coefficient of kurtosis*, was introduced in Section 5.7.

EXAMPLE 5.10. What are the first four dimensionless moments of the distribution in Example 5.1, and what are their names?

*Solution.* The moments about the mean of the distribution in Example 5.1 are

$$m_1 = 0$$
$$m_2 = 1.35$$
$$m_3 = -0.15$$
$$m_4 = 5.21$$
$$\sigma = \sqrt{m_2} = \sqrt{1.35} = 1.16$$

The dimensionless moments of the distribution are

$$a_1 = 0$$
$$a_2 = \frac{1.35}{(1.16)^2} = \frac{1.35}{1.35} = 1$$
$$a_3 = -\frac{0.15}{(1.16)^3} = -\frac{0.15}{1.56} = -0.10$$
$$a_4 = \frac{5.21}{(1.16)^4} = \frac{5.21}{1.81} = 2.88$$

The first two dimensionless moments being always 0 and 1 respectively, have no names. The third dimensionless moment is the moment coefficient of skewness, and the fourth is the moment coefficient of kurtosis.

EXAMPLE 5.11. What are the values of the first four dimensionless moments of the normal curve.

*Solution.* The first two values are 0 and 1 respectively, as they are for all curves. Since the normal curve is symmetrical, the third dimensionless moment is 0. As stated in Section 5.7 the fourth dimensionless moment of the normal curve is 3. Hence,

$$a_1 = 0$$

$$a_2 = 1$$

$$a_3 = 0$$

$$a_4 = 3$$

**5.9. Sheppard's Corrections and Charlier's Check.** It will be remembered that in Chapter 4, Section 4.10, the corrected value of the standard deviation, $\sigma^2$, for grouped data (*Sheppard's Correction*) was

$$\sigma^2 - \frac{C^2}{12}$$

Similar corrections to higher moments can be applied. If ${}_gm_1$, ${}_gm_2$, ${}_g\mathrm{m}_3$, etc., are the moments calculated for grouped data, then the corrected moments are

$$m_1 = {}_gm_1$$

$$m_2 = {}_gm_2 - \frac{1}{12}C^2$$

$$m_3 = {}_gm_3$$

$$m_4 = {}_gm_4 - \frac{1}{2}C^2{}_gm_2 + \frac{7}{240}C^4$$

As explained in Section 4.10, these corrections will not be used in this book.

*Charlier's Check* was described in Chapter 4, Section 4.11, and can be extended to higher moments.

$$\Sigma f(u+1) = \Sigma fu + n$$

$$\Sigma f(u+1)^2 = \Sigma f(u)^2 + 2\Sigma fu + n$$

$$\Sigma f(u+1)^3 = \Sigma f(u)^3 + 3\Sigma f(u)^2 + 3\Sigma fu + n$$

$$\Sigma f(u+1)^4 = \Sigma f(u)^4 + 4\Sigma f(u)^3 + 6\Sigma f(u)^2 + 4\Sigma fu + n$$

Since higher moments are usually calculated by data processing equipment, this check is rarely employed in practice.

**5.10. Notation—Universe and Samples.** When it is required to distinguish between functions relating to the *total population* or *universe*, and functions relating to a *sample*, Greek letters are used for the universe and Latin letters for the sample. Thus, the moments of the universe are represented by $\mu$ (Greek mu) or $\mu'$ and skewness and kurtosis of the universe by $\alpha_3$ and $\alpha_4$ (Greek alpha). In the same way $\sigma$ (Greek sigma) is the standard deviation of the universe and $s$ the standard deviation of the sample.

## Problems

Moments higher than the first, which establishes the mean, have no significance for very small volumes of data. However, so as to provide tests of the principles involved in the calculation of higher moments without presenting the student with long laborious calculations, some of the following problems are based on very limited data.

**Problem 5.1.** The ages of children attending a party are

2, 3, 3, 4, 5

What are the first four moments of the ages?

**Problem 5.2.** What are the first four moments about the mean of the data in Problem 5.1, calculated directly from the mean?

**Problem 5.3.** Calculate the first four moments about the mean of the data in Problem 5.1, using the relationships between the moments about the mean and the moments about the origin.

**Problem 5.4.** What are the moment coefficients of skewness and of kurtosis of these data?

**Problem 5.5.** Is the distribution in Problem 5.1 skewed to the left or the right? Is the distribution platykurtic or leptokurtic?

**Problem 5.6.** Calculate the first four moments of the data in Problem 5.1, using 3 as origin. Use these results to calculate the first four moments about the mean.

**Problem 5.7.** In a study of road accidents, the following was the distribution for number of vehicles involved in each accident:

| Number of Vehicles | No. of Accident Involvements |
|---|---|
| 1 | 2241 |
| 2 | 3272 |
| 3 | 264 |
| 4 | 47 |
| 5 | 6 |
| 6 or more | 3 |

Calculate the first four moments about the mean of the distribution.

**Problem 5.8.** What are the mean and standard deviation of the data in Problem 5.7? What are the moment coefficients of skewness and kurtosis?

**Problem 5.9.** The distribution of days absent during a year among the employees of a small firm are as follows:

| No. of Days Absent | Number of Employees |
|---|---|
| 0– 4 | 5 |
| 5– 9 | 10 |
| 10–14 | 8 |
| 15–19 | 1 |
| 20–24 | 0 |
| 25–29 | 1 |
| | 25 |

Calculate the first four moments of this distribution about the origin.

**Problem 5.10.** Calculate the first four moments about the mean for the distribution in Problem 5.9.

**Problem 5.11.** Calculate the first four moments about the mean for the data in Problem 5.9, using an arbitrary origin of 7 and units of 5.

**Problem 5.12.** What are the mean and the standard deviation of the data in Problem 5.9?

**Problem 5.13.** What are the moment coefficients of skewness and kurtosis of the data in Problem 5.9?

**Problem 5.14.** Calculate Pearson's first and second coefficients of skewness for the distribution in Example 3.6. Is the distribution skewed to the left or to the right?

**Problem 5.15.** What is the percentile measure of kurtosis of the distribution in Example 3.9?

**Problem 5.16.** What are the first four dimensionless moments of the distribution in Problem 5.1?

**Problem 5.17.** What are the first four dimensionless moments of the distribution in Problem 5.7?

**Problem 5.18.** What are the first four dimensionless moments of the distribution in Problem 5.9?

**Problem 5.19.** What are the first four moments about the mean of the data in Problem 5.9 with Sheppard's Corrections?

**Problem 5.20.** Apply Charlier's Check of the calculation of the first three moments of the data in Example 5.4.

## Solutions

**Problem 5.1.** The expression for the $r$th moment is

$$\frac{\sum_{i=0}^{n} \bar{x}_i^r}{n}$$

$$\text{First moment} = \frac{2 + 3 + 3 + 4 + 5}{5} = \frac{17}{5} = 3.4 = (m_1')$$

$$\text{Second moment} = \frac{2^2 + 3^2 + 3^2 + 4^2 + 5^2}{5}$$

$$= \frac{4 + 9 + 9 + 16 + 25}{5} = \frac{63}{5} = 12.6 = (m_2')$$

$$\text{Third moment} = \frac{2^3 + 3^3 + 3^3 + 4^3 + 5^3}{5}$$

$$= \frac{8 + 27 + 27 + 64 + 125}{5} = \frac{251}{5}$$

$$= 50.2 = (m_3')$$

$$\text{Fourth moment} = \frac{2^4 + 3^4 + 3^4 + 4^4 + 5^4}{5}$$

$$= \frac{16 + 81 + 81 + 256 + 625}{5} = \frac{1059}{5}$$

$$= 211.8 = (m_4')$$

**Problem 5.2.** The mean of the data is 3.4, and the first moment about the mean ($m_1$) is 0. The second moment about the mean ($m_2$) is

$$m_2 = \frac{(-1.4)^2 + (-.4)^2 + (-.4)^2 + (.6)^2 + (1.6)^2}{5}$$

$$= \frac{1.96 + 0.16 + 0.16 + 0.36 + 2.56}{5} = \frac{5.20}{5} = 1.04$$

The third moment about the mean ($m_3$) is

$$m_3 = \frac{(-1.4)^3 + (-.4)^3 + (-.4)^3 + (.6)^3 + (1.6)^3}{5}$$

$$= \frac{-2.744 - .064 - .064 + .216 + 4.096}{5} = \frac{1.440}{5} = 0.29$$

The fourth moment about the mean ($m_4$) is

$$m_4 = \frac{(-1.4)^4 + (-.4)^4 + (-.4)^4 + (.6)^4 + (1.6)^4}{5}$$

Using tables of fourth powers and rounding to two places of decimals, we have

$$m_4 = \frac{3.84 + .02 + .02 + .13 + 6.55}{5} = \frac{10.56}{5} = 2.11$$

**Problem 5.3.**

$$m_1 = 0$$

$$\begin{aligned} m_2 &= m_2' - (m_1')^2 = 12.6 - (3.4)^2 \\ &= 12.6 - 11.56 \\ &= 1.04 \end{aligned}$$

$$\begin{aligned} m_3 &= m_3' - 3m_1'm_2' + 2(m_1')^3 \\ &= 50.2 - (3 \times 3.4 \times 12.6) + 2(3.4)^3 \\ &= 50.2 - (3 \times 42.84) + (2 \times 39.3) \\ &= 50.2 - 128.5 + 78.6 \\ &= 0.3 \end{aligned}$$

$$\begin{aligned} m_4 &= m_4' - 4m_1'm_3' + 6(m_1')^2m_2' - 3(m_1')^4 \\ &= 211.8 - (4 \times 3.4 \times 50.2) + (6(3.4)^2 \times 12.6) - 3(3.4)^4 \\ &= 211.8 - 682.7 + 873.9 \; 400.9 \\ &= 2.1 \end{aligned}$$

These answers, except for the results of rounding, are the same as those calculated in the solution to Problem 5.2.

**Problem 5.4.** The moment coefficient of skewness is

$$a_3 = \frac{m_3}{\sqrt{m_2^3}} = \frac{0.3}{\sqrt{(1.04)^3}} = \frac{0.3}{1.06} = 0.3$$

The moment coefficient of kurtosis is

$$a_4 = \frac{m_4}{\sqrt{m_2^4}} = \frac{2.1}{\sqrt{1.04^4}} = \frac{2.1}{1.08} = 1.9$$

**Problem 5.5.** Since the moment coefficient of skewness is positive, the distribution is skewed to the right. Since the moment coefficient of kurtosis is less than 3, the distribution is platykurtic.

**Problem 5.6.** Using 3 as origin, the values become

$$-1,\ 0,\ 0,\ 1,\ 2$$

The moments are

$$m_1' = \frac{-1 + 1 + 2}{5} = \frac{2}{5} = .4$$

$$m_2' = \frac{1 + 1 + 4}{5} = \frac{6}{5} = 1.2$$

$$m_3' = \frac{-1 + 1 + 8}{5} = \frac{8}{5} = 1.6$$

$$m_4' = \frac{1 + 1 + 16}{5} = \frac{18}{5} = 3.6$$

The moments about the mean are

$$\begin{aligned}
m_1 &= 0 \\
m_2 &= m_2' - (m_1')^2 = 1.2 - (.4)^2 = 1.2 - .16 = 1.04 \\
m_3 &= m_3' - 3m_1'm_2' + 2(m_1')^3 \\
&= 1.6 - 3(.4)(1.2) + 2(.4)^3 \\
&= 1.6 - 1.44 + .13 = 0.3 \\
m_4 &= m_4' - 4m_1'm_3' + 6(m_1'^2)m_2' - 3(m_1')^4 \\
&= 3.6 - 4(.4)(1.6) + 6(.4)^2(1.2) - 3(.4)^4 \\
&= 3.6 - 2.56 + 1.15 - .08 = 2.1
\end{aligned}$$

It will be noted that the calculations are less laborious than those used to obtain the same results in Problems 5.2 and 5.3.

**Problem 5.7.** Using 2 as origin, the calculations are as follows. Owing to the very small size of the last group, the fact that more than 6 vehicles might have been involved in one of the three accidents may be ignored.

| d | $d^2$ | $d^3$ | $d^4$ | f | fd | $fd^2$ | $fd^3$ | $fd^4$ |
|---|---|---|---|---|---|---|---|---|
| −1 | 1 | −1 | 1 | 2241 | −2241 | 2241 | −2241 | 2241 |
| 0 | 0 | 0 | 0 | 3272 | 0 | 0 | 0 | 0 |
| 1 | 1 | 1 | 1 | 264 | 264 | 264 | 264 | 264 |
| 2 | 4 | 8 | 16 | 47 | 94 | 188 | 376 | 752 |
| 3 | 9 | 27 | 81 | 6 | 18 | 54 | 162 | 486 |
| 4 | 16 | 64 | 256 | 3 | 12 | 48 | 192 | 768 |
| | | | | 5833 | −1853 | 2795 | −1247 | 4511 |

Dividing by 5833, the total frequency, we have

$$m_1' = -.32$$
$$m_2' = .48$$
$$m_3' = -.21$$
$$m_4' = .77$$

Considerable care with signs will be needed in calculating the moments about the mean.

$$m_1 = 0$$

$$\begin{aligned} m_2 &= .48 - (-.32)^2 \\ &= .48 - .10 \\ &= .38 \end{aligned}$$

$$\begin{aligned} m_3 &= -.21 - 3(-.32)(.48) + 2(-.32)^3 \\ &= -.21 + .46 - .07 \\ &= .18 \end{aligned}$$

$$\begin{aligned} m_4 &= .77 - 4(-.32)(-.21) + 6(-.32)^2(.48) - 3(-.32)^4 \\ &= .77 - .27 + .29 - .03 \\ &= .76 \end{aligned}$$

**Problem 5.8.** The mean of the distribution is

$$2 - .32 = 1.68$$

The standard deviation is $\sqrt{m_2} = \sqrt{.38} = .62$.

The moment coefficient of skewness is

$$\frac{m_3}{\sqrt{m_2^3}} = \frac{.18}{\sqrt{.38^3}} = \frac{.18}{.23} = .8$$

The distribution is skewed to the right.

The moment coefficient of kurtosis is

$$\frac{m_4}{\sqrt{m_2^4}} = \frac{.76}{(.38)^2} = \frac{.76}{.14} = 5.4$$

The distribution is leptokurtic.

**Problem 5.9.** The midpoints of the class intervals are 2, 7, 12, 17, and the use of a unit of 5 ages does not appreciably reduce the calculation since the distance of the midpoints from the origin become .4, 1.4, 2.4, 3.4, etc. The calculations proceed as follows:

| No. of Days Absent | x | $x^2$ | $x^3$ | $x^4$ | f | fx | $fx^2$ | $fx^3$ (00) | $fx^4$ (000) |
|---|---|---|---|---|---|---|---|---|---|
| 0– 4 | 2 | 4 | 8 | 16 | 5 | 10 | 20 | 0 | 0 |
| 5– 9 | 7 | 49 | 343 | 2401 | 10 | 70 | 490 | 34 | 24 |
| 10–14 | 12 | 144 | 1728 | 20736 | 8 | 96 | 1152 | 138 | 166 |
| 15–19 | 17 | 289 | 4913 | 83521 | 1 | 17 | 289 | 49 | 84 |
| 20–24 | 22 | 484 | 10648 | 234256 | 0 | 0 | 0 | 0 | 0 |
| 25–29 | 27 | 729 | 19683 | 531441 | 1 | 27 | 729 | 197 | 531 |
| | | | | | 25 | 220 | 2680 | 418 | 805 |

The calculation of the last two columns has been simplified by the omitting of 2 and 3 digits respectively, as these digits are not significant. It must be remembered to restore these digits in the final result. Thus,

$$m_1' = 8.8$$
$$m_2' = 107$$
$$m_3' = 16.7 \times 100 = 1{,}670$$
$$m_4' = 32.2 \times 1{,}000 = 32{,}200$$

**Problem 5.10.**

$$m_1 = 0$$
$$m_2 = 107 - (8.8)^2 = 30$$
$$m_3 = 1{,}670 - 3(8.8)(107) + 2(8.8)^3 = 208$$
$$m_4 = 32{,}200 - 4(8.8)(1{,}670) + 6(8.8)^2(107) - 3(8.8)^4$$
$$= 5{,}000 \text{ (approx.)}$$

**Problem 5.11.** Using an arbitrary origin, and units of five, the calculation of the moments about the origin is considerably simplified as shown below. It must be remembered to multiply the $r$th moment by $C^r$, that is by 5, 25, 125, and 625, respectively. However, *these multiplications need not be made until the final results are obtained.*

| No. of Days Absent | u | u² | u³ | u⁴ | f | fu | fu² | fu³ | fu⁴ |
|---|---|---|---|---|---|---|---|---|---|
| 0– 4 | −1 | 1 | −1 | 1 | 5 | −5 | 5 | −5 | 5 |
| 5– 9 | 0 | 0 | 0 | 0 | 10 | 0 | 0 | 0 | 0 |
| 10–14 | 1 | 1 | 1 | 1 | 8 | 8 | 8 | 8 | 8 |
| 15–19 | 2 | 4 | 8 | 16 | 1 | 2 | 4 | 8 | 16 |
| 20–24 | 3 | 9 | 27 | 81 | 0 | 0 | 0 | 0 | 0 |
| 25–29 | 4 | 16 | 64 | 256 | 1 | 4 | 16 | 64 | 256 |
| | | | | | 25 | 9 | 33 | 75 | 285 |

$$m_1' = \frac{9}{25} \times 5 = .36 \times 5$$

$$m_2' = \frac{33}{25} \times 5^2 = 1.32 \times 5^2$$

$$m_3' = \frac{75}{25} \times 5^3 = 3 \times 5^3$$

$$m_4' = \frac{285}{25} \times 5^4 = 11.4 \times 5^4$$

The moments about the mean are

$$m_1 = 0$$

$$m_2 = [1.32 - (.36)^2] \times 5^2 = 1.19 \times 5^2 = 30$$

$$m_3 = [3 - 3(.36)(1.32) + 2(.36)^3] \times 5^3$$
$$= 1.66 \times 5^3 = 208$$

$$m_4 = [11.4 - 4(.36)(3) + 6(.36)^2(1.32) - 3(.36)^4] \times 5^4$$
$$= 8.056 \times 5^4 = 5{,}035$$

The difference in $m_4$ as calculated in this solution compared to the result obtained in the solution to Problem 5.10 is due to rounding. In each case the value of $m_4$ should be stated as 5,000.

**Problem 5.12.** The mean is $m_1'$ from the solution to Problem 5.9, or 7 (the origin) + $m_1'$ from the solution in Problem 5.11.

$$\text{Mean} = 8.8$$

$$\text{Standard Deviation} = \sqrt{m_2} = \sqrt{30} = 5.5$$

**Problem 5.13.** The moment coefficient of skewness is 1.3, and the moment coefficient of kurtosis = 5.6.

**Problem 5.14.** In the solution to Example 3.6, the mean was found to be 30.4, the median 31.3, and the mode 33.3. The standard deviation has to be calculated and this is found to be

$$\sqrt{179} = 13.4$$

Pearson's first coefficient of skewness is

$$\frac{30.4 - 33.3}{13.4} = -\frac{2.9}{13.4} = -.22$$

Pearson's second coefficient of skewness is

$$\frac{3(30.4 - 31.3)}{13.4} = \frac{-2.7}{13.4} = -.20$$

The distribution is skewed to the left.

**Problem 5.15.** The percentile measure of kurtosis is

$$\frac{Q}{P_{90} - P_{10}}$$

From the solution of Example 3.9, this is

$$\frac{\frac{1}{2}(77 - 67)}{81-65} = \frac{\frac{1}{2}(10)}{16} = \frac{5}{16} = .31$$

**Problem 5.16.** The moments about the mean of the distribution in Problem 5.1 are given in the solution of Problem 5.2.

$$m_1 = 0$$

$$m_2 = 1.04$$

$$m_3 = 0.29$$

$$m_4 = 2.11$$

$\sqrt{m_2} = \sqrt{1.04} = 1.02$, and the dimensionless moments are

$$a_1 = 0$$

$$a_2 = 1$$

$$a_3 = \frac{0.29}{(1.02)^3} = 0.27$$

$$a_4 = \frac{2.11}{(1.02)^4} = 1.95$$

**Problem 5.17.**

$$a_1 = 0$$

$$a_2 = 1$$

$$a_3 = .77$$

$$a_4 = 5.3$$

**Problem 5.18.**

$$a_1 = 0$$

$$a_2 = 1$$

$$a_3 = 1.3$$

$$a_4 = 5.6$$

**Problem 5.19.** The first four moments about the mean of the data in Problem 5.9 are

0, 30, 208, 5000

The class interval is 5. The corrected figures are:

$$m_1 = 0$$

$$m_2 = 30 - \left(\frac{1}{12} \times 25\right) = 28$$

$$m_3 = 208$$

$$m_4 = 5000 - \left(\frac{1}{2} \times 25 \times 30\right) + \left(\frac{7}{240} \times 625\right)$$

$$= 5000 - 375 + 18$$

Since 5000 was a rounded figure, the value of $m_4$ is still approximately 5000.

**Problem 5.20.** In the solution of Example 5.4, an arbitrary origin of 3 was selected. To apply Charlier's Check to these calculations, the results are now calculated with origin 2.

| d | $d^2$ | $d^3$ | f | fd | $fd^2$ | $fd^3$ |
|---|---|---|---|---|---|---|
| −1 | 1 | −1 | 1061 | −1061 | 1061 | −1061 |
| 0 | 0 | 0 | 494 | 0 | 0 | 0 |
| 1 | 1 | 1 | 232 | 232 | 232 | 232 |
| 2 | 4 | 8 | 137 | 274 | 548 | 1096 |
| 3 | 9 | 27 | 56 | 168 | 504 | 1512 |
| 4 | 16 | 64 | 29 | 116 | 464 | 1856 |
| 5 | 25 | 125 | 25 | 125 | 625 | 3125 |
| 6 | 36 | 216 | 8 | 48 | 288 | 1728 |
| 7 | 49 | 343 | 6 | 42 | 294 | 2058 |
| | | | 2048 | −56 | 4016 | 10546 |

Checking:

$$-56 = -2104 + 2048$$

$$4016 = 6176 + 2(-2104) + 2048$$

$$10546 = -3718 + 3(6176) + 3(-2104) + 2048$$

# 6

# Elementary Probability

**6.1. Introduction.** There are at least four possible methods of defining probability, which may be called briefly:

(1) The Classical definition
(2) The Frequency definition
(3) The "Fair Price" definition
(4) The "Degree of Rational Belief" definition

The *classical theory* dates back to Laplace's principle of *equally possible cases*, now more usually defined as *equally likely alternatives.* Under this definition, probability is an additive function $P(S)$ of the set $S$, in a given *a priori* space $R$. It will be noted that this is a purely conceptional mathematical theory. The term *equally likely* is difficult to interpret except in constructed problems involving dice and cards.

For most statistical studies, it is usual to consider probability as the *limiting value of the relative frequency*, when the number of observations is very large. For example, if we throw a die a very large number of times and determine the relative frequency with which each number comes up, we have a set of empirical probabilities which are valid for the particular die used, whether it is loaded or a true die.

Unfortunately, this definition, too, presents problems because, in most practical cases, conditions are not static and it is not possible to repeat a statistical study more than a limited number of times.

Whitworth, whose book *Choice and Chance* is well known, gives the following explanation of the "Fair Price" definition. "The measurement of chance (probability) may be approached *ab initio* from the consideration of the price that may reasonably be paid for 'a gain contingent on some doubtful occurrence.' And to many minds this method appears easier than any other."

The "Degree of Rational Belief" definition presupposes that we have some prior knowledge, and that we are not carrying out our experiments in a vacuum. This credibility approach is especially associated with Bayes' Theorem of *a priori probabilities.*

**6.2. Classical Probability.** In the next chapter, we shall consider *probability distributions*, which play an important role in statistical work since they provide valuable mathematical models. In order to understand these distributions some knowledge of classical probability is essential, and this chapter will provide a brief review of the theory.

If an event $E$ can occur in $a$ ways out of a total of $n$ possible equally likely ways, then the probability of the event happening is

$$p = \Pr\{E\} = \frac{a}{n} \tag{6.1}$$

The probability that the event will not happen is

$$q = \Pr\{\text{not } E\} = 1 - p = \frac{n - a}{n} \tag{6.2}$$

Pr {not $E$} may be written Pr $\{\sim E\}$ or Pr $\{\bar{E}\}$.

$$p + q = 1, \text{ and } \Pr\{E\} + \Pr\{\text{not } E\} = 1$$

EXAMPLE 6.1. What is the probability of throwing a 6 with one die? A 7 with two dice?

*Solution.* A thrown die can end up with any one of its six sides uppermost. Each is equally likely, so a single die can be thrown in 6 ways. In only one way will a six be uppermost, so that

$$p = \frac{1}{6}$$

If two dice are thrown, each die can end up in six different ways, and since the dice are independent of each other, two dice can be thrown in $6 \times 6$, or 36, different ways. These are listed below.

| 1, 1 | 2, 1 | 3, 1 | 4, 1 | 5, 1 | 6, 1 |
|---|---|---|---|---|---|
| 1, 2 | 2, 2 | 3, 2 | 4, 2 | 5, 2 | 6, 2 |
| 1, 3 | 2, 3 | 3, 3 | 4, 3 | 5, 3 | 6, 3 |
| 1, 4 | 2, 4 | 3, 4 | 4, 4 | 5, 4 | 6, 4 |
| 1, 5 | 2, 5 | 3, 5 | 4, 5 | 5, 5 | 6, 5 |
| 1, 6 | 2, 6 | 3, 6 | 4, 6 | 5, 6 | 6, 6 |

The total of the two dice will be 7 in the six ringed cases, and hence the probability of throwing 7 with two dice is

$$\frac{6}{36}, \text{ or } \frac{1}{6}$$

**6.3. Relation to Set Theory.** It is sometimes convenient to think of the different ways in which an event can occur as points in a sample space. In a given situation, the *sample space* is defined as all events which can possibly occur. If, in a sample space, there are $n$ equally likely events which can occur, then the probability of each event is $\frac{1}{n}$. All probabilities lie in the range 0 to 1, 0 indicating an impossible event and 1 a certainty.

If we have two sets in the sample space, $E_1$ and $E_2$, then

$E_1 + E_2$ or $E_1 \cup E_2$ (the union of sets $A$ and $B$) is the set of all elements belonging *either* to $E_1$ or $E_2$.

$E_1 E_2$ or $E_1 \cap E_2$ (the intersection of $A$ and $B$) is the set of all elements belonging *both* to $E_1$ and to $E_2$.

From the Venn Diagrams (Figure 6.1) we see that if the events $E_1$ and $E_2$ have no points in common or are *mutually exclusive*,

$$\Pr\{E_1 + E_2\} = \Pr\{E_1\} + \Pr\{E_2\}$$

However, if they have points in common, and are not mutually exclusive,

$$\Pr\{E_1 + E_2\} = \Pr\{E_1\} + \Pr\{E_2\} - \Pr\{E_1 E_2\} \qquad (6.3)$$

Note that for mutually exclusive events, $\Pr\{E_1 E_2\} = 0$

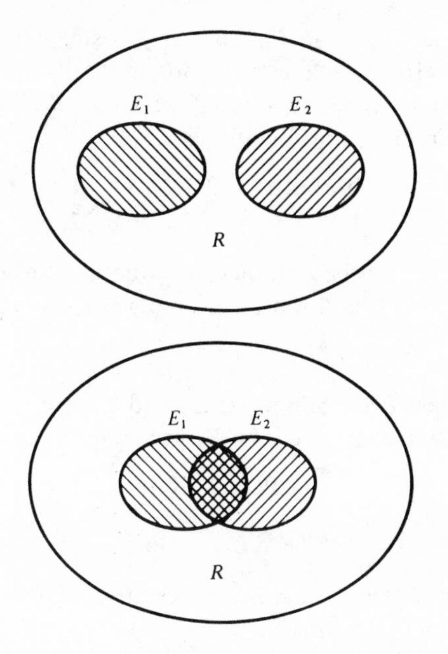


Figure 6.1. Examples of Venn diagrams.

EXAMPLE 6.2. If $E_1 + E_2$ is the sample space, what is the value of $\Pr\{E_1 + E_2\}$?
*Solution.* Since $E_1 + E_2$ is the whole sample space, $\Pr\{E_1 + E_2\} = 1$.

EXAMPLE 6.3. What is the probability of throwing *either* a double six *or* a three, on the throw of two dice?
*Solution.* From the table in the solution to Example 6.1, the probability of throwing a double six is 1/36 and the probability of throwing a three is 2/36 (since three can be 2 + 1 or 1 + 2). Hence, since these probabilities are mutually exclusive, the required probability is

$$\frac{1}{36} + \frac{2}{36} = \frac{3}{36} = \frac{1}{12}$$

EXAMPLE 6.4. What is the probability of throwing either a single 1 *or* a single 3 on the throw of two dice?
*Solution.* A single 1 can be thrown in 10 ways. (The first row and the first column in the diagram, less the first case 1 + 1 which

is not the throw of a *single* 1). Similarly, a single 3 can be thrown in 10 ways. However, these are not mutually exclusive events, since a 1 *and* a 3 can be thrown in 2 ways.

Hence, the required probability is

$$\frac{10}{36} + \frac{10}{36} - \frac{2}{36} = \frac{18}{36} = \frac{1}{2}$$

The same result can be obtained by counting those cases, in the 36 tabulated in the solution to Example 6.1, in which a single 1 or a single 3 occurs.

**6.4. Conditional Probability.** If $E_1$ and $E_2$ are two events, then the probability that $E_2$ occurs if $E_1$ occurs is

$$\Pr\{E_2 \mid E_1\}$$

If

$$\Pr\{E_2 \mid E_1\} = \Pr\{E_2\}$$

then $E_1$ and $E_2$ are *independent* events. Otherwise they are *dependent*. For *independent* events

$$\Pr\{E_1 E_2\} = \Pr\{E_1\}\Pr\{E_2\} \tag{6.4}$$

For *dependent* events

$$\Pr\{E_1 E_2\} = \Pr\{E_1\}\Pr\{E_2 \mid E_1\} \tag{6.5}$$

For three events

$$\Pr\{E_1 E_2 E_3\} = \Pr\{E_1\}\Pr\{E_2 \mid E_1\}\Pr\{E_3 \mid E_1 E_2\}$$

and this reduces to $\Pr\{E_1\}\Pr\{E_2\}\Pr\{E_3\}$ if the three events are all independent.

EXAMPLE 6.5. A bag contains two white and one black ball. What is the probability that first a white ball and then a black ball is drawn from the bag; (1) if the first ball is returned, and (2) if the first ball is not returned to the bag?

*Solution.* (1) If the first ball is returned we have two independent events. $\Pr\{E_1\}$ (the probability of drawing a white ball) is 2/3, and $\Pr\{E_2\}$ (the probability of drawing a black ball) is 1/3, and

$$\Pr\{E_1 E_2\} = \frac{2}{3} \times \frac{1}{3} = \frac{2}{9}$$

(2) If the ball is not returned, $\Pr\{E_1\}$ is still 2/3, but when the white ball has been drawn, the bag is left with one white and one black ball, so that

$$\Pr\{E_2 \mid E_1\} = \frac{1}{2}$$

$$\Pr\{E_1 E_2\} = \frac{2}{3} \times \frac{1}{2} = \frac{1}{3}$$

**6.5. Continuous Probabilities.** In the examples used to illustrate the foregoing discrete probabilities, throwing dice and drawing balls from bags have been used. Probability theory, and the formulas given, apply equally to continuous probabilities, but many of the applications require the use of calculus. Given a *probability distribution function*, such as that shown in Figure 6.2, the probability of a value lying between *a* and *b* is the area under the curve between the ordinates *a* and *b*, expressed as a ratio to the total area under the curve. The probability of an *exact value* *a* is zero.

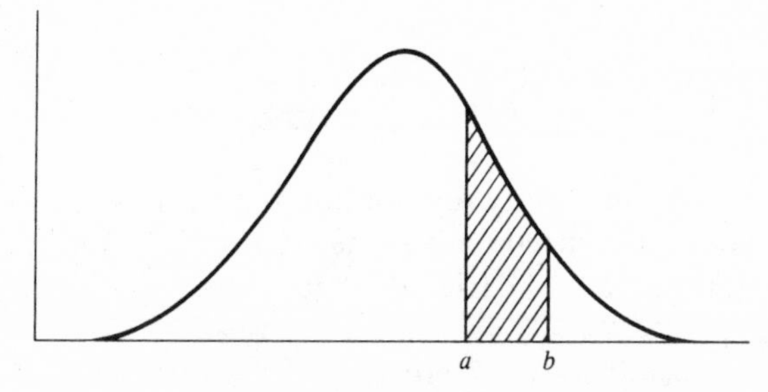


Figure 6.2. Probability distribution function.

EXAMPLE 6.6. In a study of the speeds travelled by automobiles on an expressway, it was found that the speeds of automobiles could be expressed in the form of a triangular distribution shown in Figure 6.3. No automobiles were observed travelling at less than 40 miles per hour. The distribution increased in proportion to the excess over 40 m.p.h. up to 60 m.p.h., the maximum legal speed, and then decreased steadily to 70 m.p.h., the maximum observed. What is the probability that an individual automobile

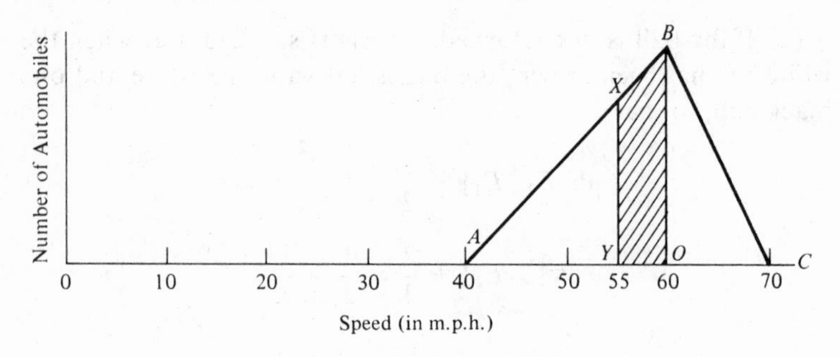


Figure 6.3. Speeds of automobiles. Probability distribution function.

(1) Is travelling at exactly 60 m.p.h.,
(2) Is travelling above the legal speed limit,
(3) Is travelling between 55 and 60 m.p.h?

*Solution.* (1) Since this is a continuous distribution, the probability of an *exact* speed of 60 m.p.h. is zero.

(2) Since the total area under the curve must be unity, the usual formula for the area of a triangle gives

$$\frac{1}{2} \times BO \times (70 - 40) = 1$$

or

$$BO = \frac{2}{30} = \frac{1}{15}$$

The area of the probability distribution representing automobiles travelling above the speed limit is the area to the right of the line $BO$, that is, the area of the triangle $BOC$. This is

$$\frac{1}{2} \times BO \times (70 - 60) = \frac{1}{2} \times \frac{1}{15} \times 10 = \frac{1}{3}$$

(3) The area of the probability distribution representing automobiles travelling between 55 and 60 m.p.h. is the shaded area in the illustration. This is

$$\frac{1}{2} \times (BO + XY) \times (60 - 55)$$

now

$$\frac{XY}{55 - 40} = \frac{BO}{60 - 40}$$

so that

$$XY = \frac{15}{20} \times \frac{1}{15} = \frac{1}{20}$$

and the required probability is

$$\frac{1}{2} \times \left(\frac{1}{15} + \frac{1}{20}\right) \times 5 = \frac{1}{2} \times \frac{7}{60} \times 5 = .29$$

**6.6. Permutations.** In many examples in classical probability, it is necessary to determine the number of *ways* in which an event can occur. For example, in Example 6.1 it was shown that two dice could be thrown in 36 ways. In more complicated problems we often want to know in how many ways can $n$ different objects be taken, $r$ at a time. This is called the number of *permutations* of $n$ things taken $r$ at a time ${}_nP_r$. Since the first thing can be chosen in $n$ ways, the second can be chosen in $n - 1$ ways (from the $n - 1$ remaining things).

$${}_nP_r = n(n - 1)(n - 2)\ldots(n - r + 1)$$

$${}_nP_r = \frac{n!}{(n - r)!} \tag{6.6}$$

where

$$n! = n(n - 1)(n - 2)\ldots 2\cdot 1$$

and is called *factorial n*.

Factorial $n$ may also be written $\underline{|n}$.

$$0! = 1 \text{ by definition.}$$

The first few values of $n!$ are

| **n** | **n!** |
|---|---|
| 1 | 1 |
| 2 | 2 |
| 3 | 6 |
| 4 | 24 |
| 5 | 120 |
| 6 | 720 |
| 7 | 5040 |

Statistical tables give values of $n!$, but after the first 10 or 15 values, these become too large to tabulate and log $n!$ is tabulated instead. Still higher values can be calculated by *Stirling's Approximation*

$$n! \approx \sqrt{2\pi n}\, n^n e^{-n} \tag{6.7}$$

where $e$ is the base of the natural logarithm and equals 2.71828 . . .

The *total number* of permutations of $n$ things is obtained by putting $r = n$ in Formula 6.6.

$$_nP_n = n! \tag{6.8}$$

and the number of permutations of $n$ things of which $r_1$ are alike, $r_2$ are alike, etc., is

$$\frac{n!}{r_1! \times r_2! \times \ldots} \tag{6.9}$$

EXAMPLE 6.7. In how many different ways can four cards be drawn from a pack of 52 cards?

*Solution.* Four cards can be drawn from 52 cards in $_{52}P_4$ ways.

$$_{52}P_4 = \frac{52!}{(52 - 4)!} = \frac{52!}{48!}$$

$$52! = 52 \times 51 \times 50 \times 49 \times 48!$$

$$\frac{52!}{48!} = 52 \times 51 \times 50 \times 49$$

$$= 6{,}497{,}400$$

Alternatively, using logarithms from statistical tables,

$$\log 52! = 67.90665$$

$$-\log 48! = -61.09391$$

$$6.81274$$

$$\text{antilog} = 6{,}497{,}000 \textit{ approx.}$$

EXAMPLE 6.8. How many different nine letter words (i.e., arrangements of letters regardless whether they have any meaning) can be formed from the word *statistic*?

*Solution. Statistic* consists of 9 letters, but $t$ appears 3 times, $s$ twice and $i$ twice. The number of arrangements is

$$\frac{9!}{3! \times 2! \times 2!}$$

$$= \frac{9 \cdot 8 \cdot 7 \cdot 6 \cdot 5 \cdot 4 \cdot 3 \cdot 2 \cdot 1}{3 \cdot 2 \cdot 2 \cdot 2} = 3 \times 7! = 3 \times 5040$$

$$= 15{,}120$$

**6.7. Combinations.** In Example 6.7, a four card selection such as Ace of Spades, King of Hearts, Four of Hearts, and Two of Clubs would be counted as different from King of Hearts, Ace of Spades, Four of Hearts, and Two of Clubs. In probability problems we often want to know the number of *combinations* of $n$ objects taken $r$ at a time, *with no regard to the order of arrangement.* This is written

$${}_nC_r \quad \text{or} \quad \binom{n}{r}$$

$$\binom{n}{r} = \frac{{}_nP_r}{r!} = \frac{n!}{r!\,(n-r)!} \tag{6.10}$$

EXAMPLE 6.9. How many different sets of five card hands can be dealt from a pack of 52 cards?
*Solution.*

$${}_{52}C_5 = \frac{52!}{5! \times 47!}$$

$$= \frac{52 \times 51 \times 50 \times 49 \times 48 \times 47!}{5 \times 4 \times 3 \times 2 \times 1 \times 47!}$$

$$= 52 \times 51 \times 10 \times 49 \times 2$$

$$= 2{,}598{,}960$$

Or using logarithms,

$$\log 52! = 67.90665$$

$$-\log 5! = -2.07918$$

$$-\log 47! = \frac{-59.41267}{6.41480}$$

$$\text{antilog} = 2{,}599{,}000 \text{ (approx.)}$$

**6.8. Mathematical Expectation.** If the probability of an event happening is $p$ and the monetary, or other measure, of amount receivable if the event happens is $S$, then the expectation is

$$S \times p$$

EXAMPLE 6.10. In a certain game, a player receives 2 points if a spade or club is drawn from a pack and 1 point if a heart is drawn. He has to give up 4 points if a diamond is drawn. What is expectation on each draw?

*Solution.* The probability of any particular suit being drawn is 1/4. The player expectation is

$$2\left(\frac{1}{4} + \frac{1}{4}\right) + 1\left(\frac{1}{4}\right) - 4\left(\frac{1}{4}\right) = 1 + \frac{1}{4} - 1 = \frac{1}{4}$$

It should be noted that expectations can be *positive or negative.*

## Problems

**Problem 6.1.** A card is drawn from a regular pack of 52 cards. $E_1$ is the event of drawing a Heart. $E_2$ is the event of drawing an Ace or a King.

Describe in words

$$(1)\ \bar{E}_1,\quad (2)\ \bar{E}_2,\quad (3)\ E_1 + E_2,$$
$$(4)\ \Pr\{\bar{E}_1 + E_2\},\quad (5)\ E_1 E_2,$$
$$(6)\ \Pr\{E_1 \bar{E}_2\}$$

**Problem 6.2.** Three coins are tossed. $E_1$ is the event of all turning up heads. $E_2$ is the event of having at least two tails.

Describe in words each of the following probabilities and calculate their values.

(1) $\Pr\{E_1\}$, (2) $\Pr\{\bar{E}_1\}$, (3) $\Pr\{E_2\}$, (4) $\Pr\{\bar{E}_2\}$,

(5) $\Pr\{E_1 + E_2\}$, (6) $\Pr\{\bar{E}_1 + \bar{E}_2\}$, (7) $\Pr\{E_1 + \bar{E}_2\}$,

(8) $\Pr\{\bar{E}_1 + E_2\}$, (9) $\Pr\{E_1 E_2\}$, (10) $\Pr\{\bar{E}_1 \bar{E}_2\}$,

(11) $\Pr\{E_1 \bar{E}_2\}$, (12) $\Pr\{\bar{E}_1 E_2\}$

**Problem 6.3.** Check the solution to Problem 6.2(5) by Formula 6.3.

$$\Pr\{E_1 + E_2\} = \Pr\{E_1\} + \Pr\{E_2\} - \Pr\{E_1 E_2\}$$

and also Problem 6.2(6), (7), and (8),

$$\Pr\{\bar{E}_1 + \bar{E}_2\}, \Pr\{E_1 + \bar{E}_2\}, \text{and } \Pr\{\bar{E}_1 + E_2\}$$

**Problem 6.4.** A bag contains 2 black and 3 white balls. $E_1$ is the event of drawing a white ball at the first drawing. $E_2$ is the event of drawing a white ball at the second drawing. The ball drawn is replaced after each drawing. Express in terms of $E_1$ and $E_2$ the following probabilities, and evaluate them.

(1) Drawing two white balls.
(2) Drawing two black balls.
(3) Drawing one white and one black ball.

**Problem 6.5.** Find the solution to Problem 6.4 if the ball drawn is not replaced.
**Problem 6.6.** If $E_1 + E_2$ is the sample space and $\Pr\{E_1\} = p_1$ and $\Pr\{E_2\} = p_2$, what is the value of $\Pr\{E_1 E_2\}$?
**Problem 6.7.** Twelve cards are numbered 1 to 12. Two cards are drawn without replacement. What is the probability that the greater of the two numbers drawn exceeds 6?
**Problem 6.8.** Three men each toss a coin simultaneously to decide who shall pay for dinner. If two heads and a tail come up, the man with the tail pays. Similarly, if two tails and a head came up, the odd man pays. If three heads or three tails come up, they play again. What is the probability that a decision as to who shall pay is reached on the third round of tossing?
**Problem 6.9.** Bag *A* contains 5 black balls and 2 red balls. Bag *B* contains 2 black balls and 3 red balls. I take a ball out of bag *A*, and without looking at it, place it in bag *B*. I now shake up bag *B* and draw out a ball. What is the probability that it is red?
**Problem 6.10.** Check the result in Problem 6.9 by calculating the probability of drawing a black ball at the second drawing.
**Problem 6.11.** A certain individual is never early for an appointment, but he may be up to 6 minutes late. The probability of him arriving at his appointment at time $t$ after the due time is proportional to $(6 - t)$ where $t$ is measured in minutes. What is the probability that (1) he will be no more than 1 minute late, (2) that he will be between 3 and 5 minutes late?
**Problem 6.12.** What are the number of ways in which 8 people may be seated at table?
**Problem 6.13.** In how many ways can a president, a vice-

president, a secretary, and a treasurer be selected for a club with 12 members?

**Problem 6.14.** In how many ways can a foursome be selected from 12 members of a golf club?

**Problem 6.15.** In how many ways can a pack of 52 playing cards be distributed into four hands of 13 cards each? (Do not evaluate your solution.)

**Problem 6.16.** A single die is thrown until a six has come up on two occasions. What is the probability that this will be on the $n$th throw?

**Problem 6.17.** If the die in Problem 6.16 is thrown until a six has come up on *three* occasions, what is the probability that this will be on the $n$th throw?

**Problem 6.18.** Two men roll a die alternately. $A$ wins if he rolls a 5 or 6 before $B$ rolls a 1, 2, or 3. $A$ rolls first. What is the probability that he wins?

**Problem 6.19.** Under the conditions of Problem 6.18, what would be $B$'s probability of winning if he rolled first?

**Problem 6.20.** $A$ wagers $B$ as follows. $A$ will draw two cards from a 52 card pack. If at least one is a spade, $B$ will pay $A$ \$2. If both are from the same suit, $A$ will pay $B$ \$5. If one of them is the King of Hearts, $B$ will pay $A$ \$10. What is $A$'s expectation?

## Solutions

**Problem 6.1.**

(1) Drawing a Spade, Club, or Diamond.
(2) Drawing a Queen, Jack, Ten, . . . , or Two.
(3) Drawing a Heart, an Ace, or a King.
(4) The probability of drawing a Spade, Club or Diamond or an Ace or King.
(5) Drawing the Ace of Hearts or the King of Hearts.
(6) The probability of drawing the Queen of Hearts, the Jack of Hearts, the ten of Hearts, . . . , to the two of Hearts.

**Problem 6.2.**

(1) Probability of $HHH = \frac{1}{8}$

(2) Probability of $HHT$, $HTT$, or $TTT = \frac{7}{8}$

(3) Probability of $HTT$ or $TTT = \frac{1}{2}$

(4) Probability of $HHH$ or $HHT = \frac{1}{2}$

(5) Probability of $HHH$, $HTT$, or $TTT = \frac{5}{8}$

(6) Probability of $HHH$, $HHT$, $HTT$, or $TTT = 1$

(7) Probability of $HHH$ or $HHT = \frac{1}{2}$

(8) Probability of $HHT$, $HTT$, or $TTT = \frac{7}{8}$

(9) Probability of $HHH$ *and* ($HTT$ or $TTT$) $= 0$

(10) Probability of $HHT = \frac{3}{8}$

(11) Probability of $HHH = \frac{1}{8}$

(12) Probability of $HTT$ or $TTT = \frac{1}{2}$

**Problem 6.3.** $\Pr\{E_1 + E_2\}$ is the answer to (5) and this gives

$$\frac{5}{8} = \frac{1}{8} + \frac{1}{2} - 0 = \frac{5}{8}$$

Similarly for the other three tests.

**Problem 6.4.** Since the ball drawn is replaced, $E_1$ and $E_2$ are independent and $\Pr\{E_1\} = \Pr\{E_2\}$.

(1) $\Pr\{E_1 E_2\} = \Pr\{E_1\}\Pr\{E_1\} = \left(\frac{3}{5}\right)^2 = \frac{9}{25}$

(2) $\Pr\{\bar{E}_1 \bar{E}_2\} = \Pr\{\bar{E}_1\}\Pr\{\bar{E}_1\} = \left(\frac{2}{5}\right)^2 = \frac{4}{25}$

(3) $\Pr\{E_1 \bar{E}_2\} + \Pr\{\bar{E}_1 E_2\} = 2\Pr\{E_1\}\Pr\{\bar{E}_1\}$

$$= 2 \times \frac{3}{5} \times \frac{2}{5} = \frac{12}{25}$$

**Problem 6.5.**

(1) $\Pr\{E_1 E_2\} = \Pr\{E_1\}\Pr\{E_2 \mid E_1\}$

$$= \frac{3}{5} \times \frac{2}{4} = \frac{3}{10}$$

$$(2)\ \Pr\{\bar{E}_1\bar{E}_2\} = \Pr\{\bar{E}_1\}\Pr\{\bar{E}_2 \mid \bar{E}_1\}$$

$$= \frac{2}{5} \times \frac{1}{4} = \frac{1}{10}$$

$$(3)\ \Pr\{E_1\bar{E}_2\} + \Pr\{\bar{E}_1E_2\} = \Pr\{E_1\}\Pr\{\bar{E}_2 \mid E_1\} + \Pr\{\bar{E}_1\}\Pr\{E_2 \mid \bar{E}_1\}$$

$$= \frac{3}{5}\cdot\frac{2}{4} + \frac{2}{5}\cdot\frac{3}{4} = \frac{3}{5}$$

**Problem 6.6.** From Equation 6.3,

$$\Pr\{E_1 + E_2\} = \Pr\{E_1\} + \Pr\{E_2\} - \Pr\{E_1E_2\}$$

Since $E_1 + E_2$ is the sample space, $\Pr\{E_1 + E_2\} = 1$

$$\Pr\{E_1E_2\} = p_1 + p_2 - 1$$

**Problem 6.7.**

$$\frac{1}{2} + \left(\frac{1}{2} \times \frac{6}{11}\right) = \frac{17}{22}$$

**Problem 6.8.**

$$\frac{1}{4} \times \frac{1}{4} \times \frac{3}{4} = \frac{3}{64}$$

**Problem 6.9.** The probability of a black ball at the first draw is 5/7. In this case bag *B* will contain 3 black and 3 red and the probability of drawing a red ball will be 1/2. The probability of a red ball at the first draw is 2/7 and the probability of a red ball at the next drawing will be 4/6.

Probability of a red ball is

$$\left(\frac{5}{7} \times \frac{1}{2}\right) + \left(\frac{2}{7} \times \frac{4}{6}\right) = \frac{15 + 8}{42} = \frac{23}{42}$$

**Problem 6.10.** The required probability is

$$\left(\frac{5}{7} \times \frac{1}{2}\right) + \left(\frac{2}{7} \times \frac{2}{6}\right) = \frac{15 + 4}{52} = \frac{19}{42}$$

The sum of the two probabilities is

$$\frac{23 + 19}{42} = 1$$

**Problem 6.11.**

$$(1)\quad \frac{1}{2}(6 + 5) \div 18 = \frac{11}{36}$$

$$(2)\quad \frac{1}{2}(3 + 1) \times 2 \div 18 = \frac{8}{36} = \frac{2}{9}$$

**Problem 6.12.**

$${}_8P_8 = 8! = 40{,}320$$

**Problem 6.13.**

$${}_{12}P_4 = \frac{12!}{8!} = 12 \times 11 \times 10 \times 9 = 11{,}880$$

**Problem 6.14.**

$$\binom{12}{4} = \frac{12!}{8! \times 4!} = \frac{12 \times 11 \times 10 \times 9}{4 \times 3 \times 2 \times 1} = 495$$

**Problem 6.15.** The 52 cards can be arranged in 52! ways, but the arrangement of the cards within each hand will not affect the final results, nor will the arrangement of the four hands. Hence, the answer is

$$\frac{52!}{13! \times 13! \times 13! \times 13! \times 4!}$$

**Problem 6.16.** The probability of a six on the $n$th throw is $1/6$. The probability of one, and only one, six in the preceeding $(n - 1)$ throws is

$${}_{n-1}P_1 \times \frac{1}{6} \times \left(\frac{5}{6}\right)^{n-2}$$

$${}_{n-1}P_1 = (n - 1)$$

and the total probability is

$$(n - 1)\,\frac{5^{n-2}}{6^n}$$

**Problem 6.17.**

$$\frac{(n - 1)(n - 2)}{2} \times \left(\frac{1}{6}\right)^3 \times \left(\frac{5}{6}\right)^{n-3}$$

**Problem 6.18.** The probability that $A$ wins on first roll is $2/6 = 1/3$. The probability that $B$ wins on his first roll is equal to the probability that $A$ does not win on first roll (2/3), times the probability that $B$ wins on his roll ($3/6 = 1/2$). Hence the probability that $B$ wins on his first roll is 1/3. The probability that neither wins on the first roll is $2/3 \times 1/2 = 1/3$. Let $A$'s total probability of winning be $P$. Then

$$P = \frac{1}{3} + \frac{1}{3}P$$

since $A$'s probability of winning, after an unsuccessful roll by each player, must be $P$. Therefore

$$\frac{2P}{3} = \frac{1}{3}$$

or

$$P = \frac{1}{2}$$

**Problem 6.19.**

$$P = \frac{1}{2} + \frac{1}{3}P$$

therefore

$$\frac{2P}{3} = \frac{1}{2}$$

or

$$P = \frac{3}{4}$$

**Problem 6.20.**

| Event | Probability | A Receives | A's Expectation |
|---|---|---|---|
| At least one Spade | $1 - \frac{39 \times 38}{52 \times 51}$ | \$ 2 | .882 |
| Both from same suit | $\frac{12}{51}$ | −5 | −1.176 |
| King of Hearts | $\frac{1}{52}$ | 10 | .192 |
| | | | − .102 |

$A$'s expectation is approximately −10 cents.

# 7

# Probability Distributions

**7.1. Importance of Probability Distributions.** The first five chapters were concerned with developing descriptive measures of empirical frequency distributions. In many aspects of statistical work, there is the need to find a mathematical model which represents an empirical distribution fairly closely. In this chapter the principal features of the more commonly used models will be developed.

**7.2. Binomial Distribution.** The *binomial* or *Bernoulli distribution* arises naturally when events depend upon a fixed probability of occurrence $p$, and when the number of trials is limited. If the probability of an event occurring at any trial is $p$, and $n$ trials are made, then from Chapter 6, the probability of *exactly* $x$ *successes* in the $n$ trials is

$$f(x) = \frac{n!}{x!(n - x)!} p^x q^{n-x} \tag{7.1}$$

where

$$q = 1 - p$$
$$x! = x(x - 1)(x - 2)(x - 3)\ldots 1$$

and

$$0! = 1$$

The main properties of the binomial distribution are:

| | |
|---|---|
| Mean | $\mu = np$ |
| Variance | $\sigma^2 = npq$ |
| Standard Deviation | $\sigma = \sqrt{npq}$ |
| Moment coefficient of skewness | $\alpha_3 = \dfrac{q - p}{\sqrt{npq}}$ |

Moment coefficient of kurtosis $\alpha_4 = 3 + \dfrac{1 - 6pq}{npq}$

When $p = q$, this distribution is symmetrical and $\alpha_3 = 0$.

EXAMPLE 7.1. If a coin is tossed 6 times, what is the probability of 0, 1, 2, 3, 4, 5, and 6 heads occurring?
*Solution.* (Using Formula 7.1)

$n = 6, p = \frac{1}{2}, q = \frac{1}{2}$, and $x$ equals successively, 0, 1, 2, . . . , 6

Note since

$$p = q = \frac{1}{2}, \qquad p^x q^{6-x} = \frac{1}{64}$$

for all values of $x$.

| x | x! | (n − x)! | n! | $\frac{n!}{x!(n-x)!}$ | f(x) |
|---|---|---|---|---|---|
| 0 | 1 | 720 | 720 | 1 | 1/64 |
| 1 | 1 | 120 | 720 | 6 | 6/64 |
| 2 | 2 | 24 | 720 | 15 | 15/64 |
| 3 | 6 | 6 | 720 | 20 | 20/64 |
| 4 | 24 | 2 | 720 | 15 | 15/64 |
| 5 | 120 | 1 | 720 | 6 | 6/64 |
| 6 | 720 | 1 | 720 | 1 | 1/64 |

It will be noted that the last column adds up to unity, which must be so, since we have included all possible outcomes of tossing the coin six times.

EXAMPLE 7.2. Six coins are tossed 256 times and the number of times 0, 1, 2, etc., heads occur is exactly equal to the probabilities, i.e.,

| No. of Heads | Frequency |
|---|---|
| 0 | 4 |
| 1 | 24 |
| 2 | 60 |
| 3 | 80 |
| 4 | 60 |
| 5 | 24 |
| 6 | 4 |
| | 256 |

Calculate by the methods of Chapters 2, 4, and 5, the five properties listed in Section 7.2 and show that these are the same as the theoretical values for a binomial distribution with $p = 1/2$ and $n = 6$.

*Solution.* Use 3 as the arbitrary origin.

| d | d² | d³ | d⁴ | f | fd | fd² | fd³ | fd⁴ |
|---|---|---|---|---|---|---|---|---|
| −3 | 9 | −27 | 81 | 4 | −12 | 36 | −108 | 324 |
| −2 | 4 | −8 | 16 | 24 | −48 | 96 | −192 | 384 |
| −1 | 1 | −1 | 1 | 60 | −60 | 60 | −60 | 60 |
| 0 | 0 | 0 | 0 | 80 | 0 | 0 | 0 | 0 |
| 1 | 1 | 1 | 1 | 60 | 60 | 60 | 60 | 60 |
| 2 | 4 | 8 | 16 | 24 | 48 | 96 | 192 | 384 |
| 3 | 9 | 27 | 81 | 4 | 12 | 36 | 108 | 324 |
| | | | | 256 | 0 | 384 | 0 | 1536 |

The mean is seen to be 3, the origin selected, and hence,

$$m_1 = 0$$
$$m_2 = 1.5$$
$$m_3 = 0$$
$$m_4 = 6.0$$

$$\text{Mean} = 3$$
$$\text{Variance} = m_2 = 1.5$$
$$\text{Standard Deviation} = \sqrt{m_2} = \sqrt{1.5} = 1.22$$

$$\text{Moment coefficient of skewness} = \frac{m_3}{\sqrt{m_2^3}} = \frac{0}{\sqrt{1.5^3}} = 0$$

$$\text{Moment coefficient of kurtosis} = \frac{m_4}{\sqrt{m_2^4}} = \frac{6.0}{(1.5)^2} = 2.67$$

From the formulas in Section 7.2, putting $p = q = 1/2$, and $n = 6$,

Mean = 3, Variance = 1.5, Standard Deviation = 1.22

Moment coefficient of skewness = 0

$$\text{Moment coefficient of kurtosis} = 3 + \frac{1 - 1.5}{1.5}$$
$$= 3 - \frac{.5}{1.5}$$
$$= 2.67$$

EXAMPLE 7.3. Three dice are rolled. What is the probability of 0, 1, 2, and 3 sixes?
*Solution.* Using Formula 7.1, $n = 3, p = 1/6, q = 5/6$.

| x | x! | n! | (n − x)! | $\frac{n!}{x!(n-x)!}$ | $p^x q^{n-x}$ | f(x) |
|---|---|---|---|---|---|---|
| 0 | 1 | 6 | 6 | 1 | $125 \div 6^3$ | $125 \div 6^3$ |
| 1 | 1 | 6 | 2 | 3 | $25 \div 6^3$ | $75 \div 6^3$ |
| 2 | 2 | 6 | 1 | 3 | $5 \div 6^3$ | $15 \div 6^3$ |
| 3 | 6 | 6 | 1 | 1 | $1 \div 6^3$ | $1 \div 6^3$ |
| | | | | | | $216 \div 6^3 = 1$ |

The probabilities of 0, 1, 2, and 3 sixes are

$$\frac{125}{216}, \frac{75}{216}, \frac{15}{216}, \text{ and } \frac{1}{216}, \text{ respectively.}$$

EXAMPLE 7.4. Three dice are rolled a large number of times, and the relative frequencies of 0, 1, 2, and 3 sixes are found to agree with the probabilities calculated in the previous example. Calculate the properties of the distributions listed in Section 7.2, and show that they are the same as the theoretical values of a binomial distribution with $p = 1/6$ and $n = 3$.
*Solution.*

| x | $x^2$ | $x^3$ | $x^4$ | f | fx | $fx^2$ | $fx^3$ | $fx^4$ |
|---|---|---|---|---|---|---|---|---|
| 0 | 0 | 0 | 0 | 125 | 0 | 0 | 0 | 0 |
| 1 | 1 | 1 | 1 | 75 | 75 | 75 | 75 | 75 |
| 2 | 4 | 8 | 16 | 15 | 30 | 60 | 120 | 240 |
| 3 | 9 | 27 | 81 | 1 | 3 | 9 | 27 | 81 |
| | | | | 216 | 108 | 144 | 222 | 396 |

$$m_1' = .5$$
$$m_2' = .667$$
$$m_3' = 1.028$$
$$m_4' = 1.833$$

$$m_1 = 0$$
$$m_2 = .667 - .25 = .417$$
$$m_3 = 1.028 - 3(.5)(.667) + 2(.5)^3 = .278$$
$$m_4 = 1.833 - 4(.5)(1.028) + 6(.5)^2(.667) - 3(.5)^4 = .589$$

$$\text{Mean} = .5$$

$$\text{Variance} = .417$$
$$\text{Standard Deviation} = .646$$
$$\text{Moment coefficient of skewness} = \frac{.278}{(.417)^{3/2}} = 1.03$$
$$\text{Moment coefficient of kurtosis} = \frac{.589}{(.417)^2} = 3.4$$

From the formulas in Section 7.2, putting $p = 1/6$, $q = 5/6$ and $n = 3$.

$$\text{Mean} = .5, \quad \text{Variance} = .417, \quad \text{Standard Deviation} = .646$$
$$\text{Moment coefficient of skewness} = \frac{4}{6} \times \frac{1}{.646} = 1.03$$
$$\text{Moment coefficient of kurtosis} = 3 + \frac{36 - 30}{15} = 3 + \frac{6}{15} = 3.4$$

**7.3. Applications of the Binomial Distribution.** The binomial distribution provides a suitable model for very many statistical distributions which occur in nature, economics, business, psycho-

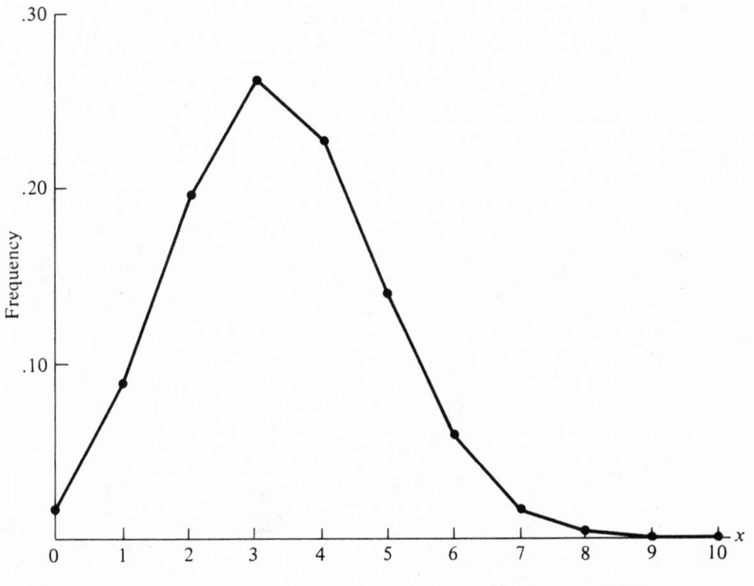


Figure 7.1. Binomial distribution. ($n = 10, p = 1/3$).

logical and educational testing, etc. The binomial is a discrete distribution, and when $n$ becomes large it approaches the normal distribution (provided neither $p$ or $q$ are close to zero). A typical distribution is shown in Figure 7.1.

**7.4. The Poisson Distribution.** The binomial distribution is not particularly appropriate to statistical studies where an event can occur more than once, for example, automobile accident experience over a *period of time* where a driver may have more than one accident in a year. In this case, the *Poisson distribution* provides a more appropriate model. This distribution is defined by the formula

$$f(x) = \frac{\lambda^x e^{-\lambda}}{x!} \tag{7.2}$$

where $x$ takes the discrete values 0, 1, 2, 3, etc. and $e$ is the base of the natural logarithm (2.71828 . . .). When $\lambda$ is small, the distribution is reversed $J$ shaped, but when $\lambda$ is large, the curve is not dissimilar to the binomial. Examples for $\lambda = .5$ and $\lambda = 4$ are shown in Figures 7.2 and 7.3.

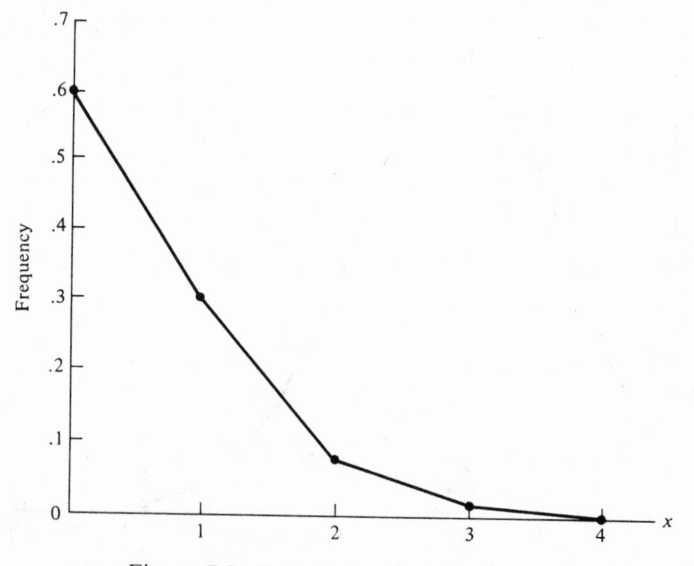


Figure 7.2. Poisson distribution. ($\lambda = .5$).

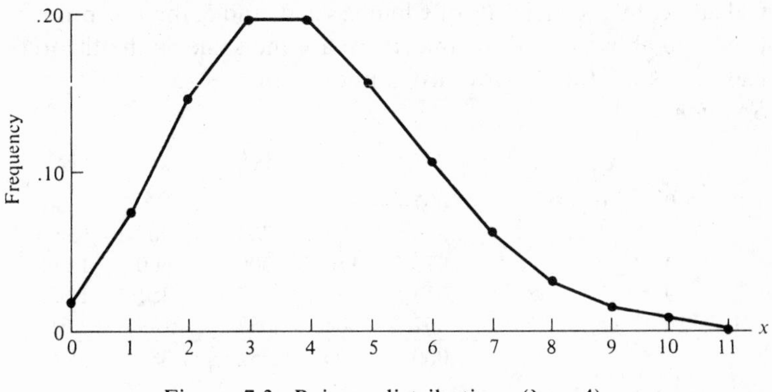


Figure 7.3. Poisson distribution. ($\lambda = 4$).

### 7.5. Properties of the Poisson Distribution

| | |
|---|---|
| Mean | $\mu = \lambda$ |
| Variance | $\sigma^2 = \lambda$ |
| Standard Deviation | $\sigma = \sqrt{\lambda}$ |
| Moment coefficient of skewness | $\alpha_3 = \dfrac{1}{\sqrt{\lambda}}$ |
| Moment coefficient of kurtosis | $\alpha_4 = 3 + \dfrac{1}{\lambda}$ |

If in the binomial distribution, $np = \lambda$ and $n$ approaches infinity, then to keep $np$ finite ($= \lambda$), $p$ must approach zero and $q$, unity. Under these conditions, the properties of the binomial distribution listed earlier approach those of the Poisson distribution. The Poisson distribution is the limit of the binomial distribution as $n \to \infty$ and $p \to 0$.

EXAMPLE 7.5. The Poisson distribution in Figure 7.2 has the following values:

| x | f(x) |
|---|---|
| 0 | .607 |
| 1 | .303 |
| 2 | .075 |
| 3 | .013 |
| 4 | .002 |
| | 1.000 |

Calculate, by the methods of Chapters 2, 4, and 5, the five properties listed above and show that these are the same as the theoretical values for the Poisson distribution with $\lambda = .5$.
*Solution.*

| x | $x^2$ | $x^3$ | $x^4$ | f | fx | $fx^2$ | $fx^3$ | $fx^4$ |
|---|---|---|---|---|---|---|---|---|
| 0 | 0 | 0 | 0 | .607 | 0 | 0 | 0 | 0 |
| 1 | 1 | 1 | 1 | .303 | .303 | .303 | .303 | .303 |
| 2 | 4 | 8 | 16 | .075 | .150 | .300 | .600 | 1.200 |
| 3 | 9 | 27 | 81 | .013 | .039 | .117 | .351 | 1.053 |
| 4 | 16 | 64 | 256 | .002 | .008 | .032 | .128 | .512 |
| | | | | 1.000 | .500 | .752 | 1.382 | 3.068 |

$$m_1' = .500$$
$$m_2' = .752$$
$$m_3' = 1.382$$
$$m_4' = 3.068$$

Therefore

$$\begin{aligned} m_1 &= 0 \\ m_2 &= .752 - .250 = .5 \\ m_3 &= 1.382 - 3(.5)(.752) + 2(.5)^3 \\ &= 1.382 - 1.128 + .25 = .5 \\ m_4 &= 3.068 - 4(.5)(1.382) + 6(.5)^2(.752) - 3(.5)^4 \\ &= 3.068 - 2.764 + 1.128 - .188 \\ &= 1.24 \end{aligned}$$

Hence,

Mean = .5, Variance = .5, Standard Deviation = .71

Moment coefficient of skewness = $.5 \div .35 = 1.4$
Moment coefficient of kurtosis = $1.24 \div .25 = 5$

Substituting .5 for $\lambda$ in the formulas, the calculated values are

Mean = .5, Variance = .5 Standard Deviation = .71

$$\text{Moment coefficient of skewness} = \frac{1}{.71} = 1.4$$

$$\text{Moment coefficient of kurtosis} = 3 + \frac{1}{.5} = 5$$

EXAMPLE 7.6. Calculate the Poisson distribution frequencies for $x = 0, 1, 2, 3$, and 5, when $\lambda = .1$.
*Solution.* First calculate the value of $e^{-\lambda}$.

$$\log e = 0.43429$$
$$.1 \log e = 0.04343$$
$$-.1 \log e = 9.95657 - 10$$
$$e^{-.1} = .9048$$

| x | $\lambda^x$ | $e^{-\lambda}$ | x! | f |
|---|---|---|---|---|
| 0 | 1 | .9048 | 1 | .9048 |
| 1 | .1 | .9048 | 1 | .0905 |
| 2 | .01 | .9048 | 2 | .0045 |
| 3 | .001 | .9048 | 6 | .0002 |
| 5 | .00001 | .9048 | 120 | .0000 |

Note, if $\lambda = .1$, the distribution is a steep reversed *J* curve.

EXAMPLE 7.7. Among the employees of a certain large organization, it is established that, on the average, 2 die each month. Assuming that the distribution of deaths by month follows the Poisson law, what is the probability of 0 deaths in a particular month?
*Solution.* The mean of the Poisson distribution is $\lambda$. Therefore,

$$\lambda = 2$$

The probability of 0 deaths is $f(0)$ which, from Formula 7.2, equals

$$\frac{\lambda^0 e^{-\lambda}}{0!}$$

Now, $\lambda^0 = 1$, $e^{-\lambda} = e^{-2}$, and $0! = 1$ (see Section 7.2). The probability of 0 deaths equals

$$e^{-2} = .135$$

**7.6. Normal Distribution.** The *normal distribution* is one of the most important probability functions. Unlike the binomial and the Poisson distributions, it is *continuous* and always *symmetrical*. The normal distribution is the limiting form of the binomial distribution if $n$ is large, provided neither $p$ nor $q$ are close to zero.

The normal distribution is also called the *normal curve* and the *Gaussian distribution.*

The formula for the normal distribution is

$$y = \frac{1}{\sigma\sqrt{2\pi}} e^{-(x - \mu)^2/2\sigma^2} \tag{7.3}$$

where $y$ is the frequency, $\mu$ is the mean, and $\sigma$ the standard deviation. $\pi$ and $e$ have the values 3.14159 . . . , and 2.71828 . . . , respectively.

When expressed in terms of standard units, where

$$z = \frac{x - \mu}{\sigma}$$

the formula becomes

$$Y = \frac{1}{\sqrt{2\pi}} e^{-z^2/2} \tag{7.4}$$

This is the *standard form* and is the equation when the mean is 0 and the variance is 1. The normal curve is shown in Figure 7.4.

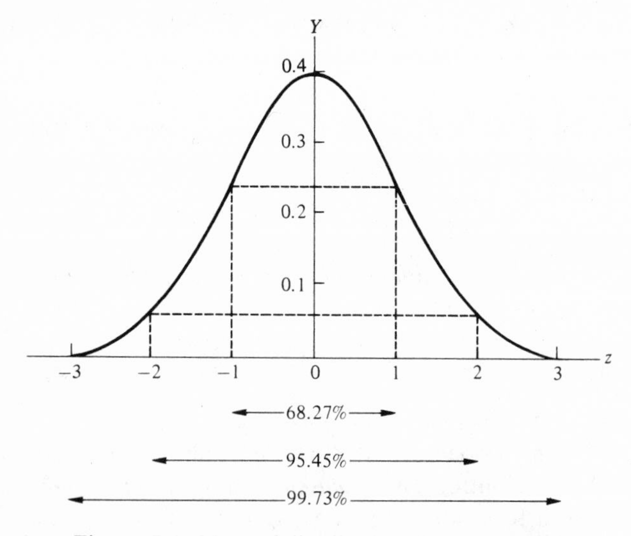


Figure 7.4. Normal distribution (Normal curve).

## 7.7. Properties of the Normal Distribution.

| | Formula 7.3 | Formula 7.4 |
|---|---|---|
| Mean | $\mu$ | 0 |
| Variance | $\sigma^2$ | 1 |
| Standard Deviation | $\sigma$ | 1 |
| Moment coefficient of skewness | 0 | 0 |
| Moment coefficient of kurtosis | 3 | 3 |

Using Formula 7.4, the height of the mean is $1/\sqrt{2\pi} = .39894$. Since the curve is symmetrical, the mean, mode, and median all coincide and

$$\text{Mean Deviation} = \sigma\sqrt{\frac{2}{\pi}} = 0.7979\sigma$$

Further, 68.27% of the total frequency lies within the range $-\sigma$ to $\sigma$; 95.45% within the range $-2\sigma$ to $2\sigma$; and 99.73% within the range $-3\sigma$ to $3\sigma$.

Since the curve is the limit of the *binomial distribution* when $n$ is large, it provides a model for numerous distributions experienced in all types of statistical work. It can be used as an approximation to the binomial if both $np$ and $nq$ are greater than 5.

EXAMPLE 7.8. Calculate the values of $Y$ if the standard form of the normal distribution corresponding to $z = -3, -2, -1, -\frac{1}{2}, 0, \frac{1}{2}, 1, 2,$ and 3.

*Solution.* Since the distribution is symmetrical, the values of $Y$ for negative values of $z$ will be the same as the values for positive values of $z$ and only the latter need be calculated. From Formula 7.4,

$$\log Y = \log\frac{1}{\sqrt{2\pi}} - \frac{1}{2}z^2\log e$$

The value of $1/\sqrt{2\pi}$ can be found in statistical tables or calculated directly; it is 0.39894. Log $e$ is given in both mathematical and statistical tables. Its value is 0.43429.

| z | $\frac{1}{2}z^2$ | $\frac{1}{2}z^2 \log e$ | $-\frac{1}{2}z^2 \log e$ | $\log \frac{1}{\sqrt{2\pi}}$ | log Y | Y |
|---|---|---|---|---|---|---|
| 0 | 0 | 0 | 0 | 9.6009-10 | 9.6009-10 | .399 |
| $\frac{1}{2}$ | .125 | 0.0543 | 9.9457-10 | 9.6009-10 | 9.5466-10 | .352 |
| 1 | .5 | 0.2172 | 9.7828-10 | 9.6009-10 | 9.3837-10 | .242 |
| 2 | 2.0 | 0.8686 | 9.1314-10 | 9.6009-10 | 8.7323-10 | .054 |
| 3 | 4.5 | 1.9543 | 8.0457-10 | 9.6009-10 | 7.6466-10 | .004 |

The values of $Y$ for the required values of $z$ are:

| z | $-3$ | $-2$ | $-1$ | $-\frac{1}{2}$ | 0 | $\frac{1}{2}$ | 1 | 2 | 3 |
|---|---|---|---|---|---|---|---|---|---|
| Y | .004 | .054 | .242 | .352 | .399 | .352 | .242 | .054 | .004 |

**7.8. Areas and Ordinates of the Standard Normal Curve.** To assist in comparing empirical distributions with the normal curve, and to help in drawing conclusions from such distributions, tables have been prepared which give the areas under the *standard normal curve* between the $y$ axis and the parallel line corresponding to any value of $z$. Tables are also available which give the heights of the $y$ ordinates corresponding to any value of $z$. Since the area under the total normal curve is unity, the maximum value in the table of areas will be .5000 (See Figure 7.5). A table of areas is given in Table 7.1.

Tables of ordinates may be calculated to agree with a total area of unity, in which case, the ordinate corresponding to $z = 0$ is

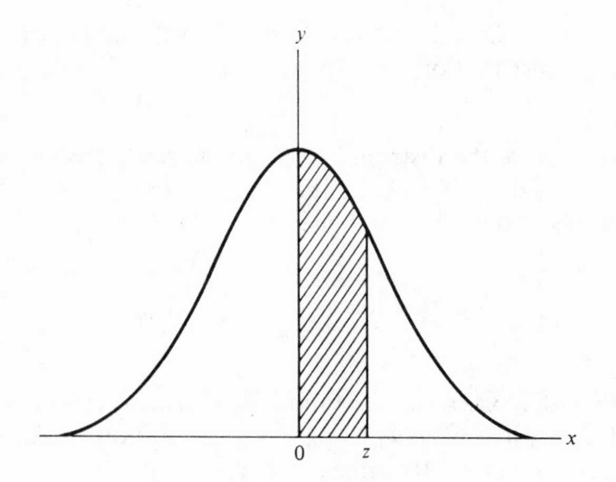


Figure 7.5. Area under the Normal curve.

## TABLE 7.1
### AREA UNDER STANDARD NORMAL CURVE

| $\frac{x}{\sigma}$ | .00 | .01 | .02 | .03 | .04 | .05 | .06 | .07 | .08 | .09 |
|---|---|---|---|---|---|---|---|---|---|---|
| 0.0 | .0000 | .0040 | .0080 | .0120 | .0159 | .0199 | .0239 | .0279 | .0319 | .0359 |
| 0.1 | .0398 | .0438 | .0478 | .0517 | .0557 | .0596 | .0636 | .0675 | .0714 | .0753 |
| 0.2 | .0793 | .0832 | .0871 | .0910 | .0948 | .0987 | .1026 | .1064 | .1103 | .1141 |
| 0.3 | .1179 | .1217 | .1255 | .1293 | .1331 | .1368 | .1406 | .1443 | .1480 | .1517 |
| 0.4 | .1554 | .1591 | .1628 | .1664 | .1700 | .1736 | .1772 | .1808 | .1844 | .1879 |
| 0.5 | .1915 | .1950 | .1985 | .2019 | .2054 | .2088 | .2123 | .2157 | .2190 | .2224 |
| 0.6 | .2257 | .2291 | .2324 | .2357 | .2389 | .2422 | .2454 | .2486 | .2518 | .2549 |
| 0.7 | .2580 | .2612 | .2642 | .2673 | .2704 | .2734 | .2764 | .2794 | .2823 | .2852 |
| 0.8 | .2881 | .2910 | .2939 | .2967 | .2995 | .3023 | .3051 | .3078 | .3106 | .3133 |
| 0.9 | .3159 | .3186 | .3212 | .3238 | .3264 | .3289 | .3315 | .3340 | .3365 | .3389 |
| 1.0 | .3413 | .3438 | .3461 | .3485 | .3508 | .3531 | .3554 | .3577 | .3599 | .3621 |
| 1.1 | .3643 | .3665 | .3686 | .3708 | .3729 | .3749 | .3770 | .3790 | .3810 | .3830 |
| 1.2 | .3849 | .3869 | .3888 | .3907 | .3925 | .3944 | .3962 | .3980 | .3997 | .4015 |
| 1.3 | .4032 | .4049 | .4066 | .4082 | .4099 | .4115 | .4131 | .4147 | .4162 | .4177 |
| 1.4 | .4192 | .4207 | .4222 | .4236 | .4251 | .4265 | .4279 | .4292 | .4306 | .4319 |
| 1.5 | .4332 | .4345 | .4357 | .4370 | .4382 | .4394 | .4406 | .4418 | .4430 | .4441 |
| 1.6 | .4452 | .4463 | .4474 | .4485 | .4495 | .4505 | .4515 | .4525 | .4535 | .4545 |
| 1.7 | .4554 | .4564 | .4573 | .4582 | .4591 | .4599 | .4608 | .4616 | .4625 | .4633 |
| 1.8 | .4641 | .4649 | .4656 | .4664 | .4671 | .4678 | .4686 | .4693 | .4699 | .4706 |
| 1.9 | .4713 | .4719 | .4726 | .4732 | .4738 | .4744 | .4750 | .4756 | .4762 | .4767 |
| 2.0 | .4773 | .4778 | .4783 | .4788 | .4793 | .4798 | .4803 | .4808 | .4812 | .4817 |
| 2.1 | .4821 | .4826 | .4830 | .4834 | .4838 | .4842 | .4846 | .4850 | .4854 | .4857 |
| 2.2 | .4861 | .4865 | .4868 | .4871 | .4875 | .4878 | .4881 | .4884 | .4887 | .4890 |
| 2.3 | .4893 | .4896 | .4898 | .4901 | .4904 | .4906 | .4909 | .4911 | .4913 | .4916 |
| 2.4 | .4918 | .4920 | .4922 | .4925 | .4927 | .4929 | .4931 | .4932 | .4934 | .4936 |
| 2.5 | .4938 | .4940 | .4941 | .4943 | .4945 | .4946 | .4948 | .4949 | .4951 | .4952 |
| 2.6 | .4953 | .4955 | .4956 | .4957 | .4959 | .4960 | .4961 | .4962 | .4963 | .4964 |
| 2.7 | .4965 | .4966 | .4967 | .4968 | .4969 | .4970 | .4971 | .4972 | .4973 | .4974 |
| 2.8 | .4974 | .4975 | .4976 | .4977 | .4977 | .4978 | .4979 | .4980 | .4980 | .4981 |
| 2.9 | .4981 | .4982 | .4983 | .4983 | .4984 | .4984 | .4985 | .4985 | .4986 | .4986 |
| 3.0 | .49865 | .4987 | .4987 | .4988 | .4988 | .4989 | .4989 | .4989 | .4990 | .4990 |
| 3.1 | .49903 | .4991 | .4991 | .4991 | .4992 | .4992 | .4992 | .4993 | .4993 | .4993 |
| 3.2 | .49931 | | | | | | | | | |
| 3.3 | .49952 | | | | | | | | | |

TABLE 7.1 (continued)

| $\frac{x}{\sigma}$ | .00 | .01 | .02 | .03 | .04 | .05 | .06 | .07 | .08 | .09 |
|---|---|---|---|---|---|---|---|---|---|---|
| 3.4 | .49966 | | | | | | | | | |
| 3.5 | .49977 | | | | | | | | | |
| 3.6 | .49984 | | | | | | | | | |
| 3.7 | .49989 | | | | | | | | | |
| 3.8 | .49993 | | | | | | | | | |
| 3.9 | .49995 | | | | | | | | | |
| 4.0 | .49997 | | | | | | | | | |

.3989; or this ordinate may be taken as unity and other ordinates calculated proportionately. A table in the former scale is given in Table 7.2.

EXAMPLE 7.9. What is the area under the standard normal curve between ordinates minus 0.5 and plus 1.54? What is the area between $-0.5\sigma$ and $+1.54\sigma$ of the normal curve with mean of 0, height of mean 2, and standard deviation 2?

*Solution.* Since the curve is symmetrical, the area between $-0.5$ and 0 is equal to the area between 0 and $+0.5$. Hence, the area required is the sum of the area between 0 and 0.5 and the area between 0 and 1.54. From Table 7.1, this area is $.1915 + .4382 = .6297$.

The height of the mean of the *standard* normal curve is .3989. If the value of the height of the mean is increased to 2, all areas will be increased in the ratio

$$\frac{2}{.3989} = 5.01$$

Further, the value of the standard deviation of the standard normal curve is 1. If the value of the standard deviation is increased to 2, all areas will be increased in the ratio $2/1 = 2$. Therefore, the area required is

$$.6297 \times 5.01 \times 2 = 6.31$$

EXAMPLE 7.10. If the mean of a normal distribution is 7, the height of the mean is 4.9 and the standard deviation is 4, find the area under the curve between 0 and 10.

TABLE 7.2

ORDINATES (Y) OF A STANDARD NORMAL CURVE AT $Z$

| $z=\frac{x}{\sigma}$ | .00 | .01 | .02 | .03 | .04 | .05 | .06 | .07 | .08 | .09 |
|---|---|---|---|---|---|---|---|---|---|---|
| 0.0 | .3989 | .3989 | .3989 | .3988 | .3986 | .3984 | .3982 | .3980 | .3977 | .3973 |
| 0.1 | .3970 | .3965 | .3961 | .3956 | .3951 | .3945 | .3939 | .3932 | .3925 | .3918 |
| 0.2 | .3910 | .3902 | .3894 | .3885 | .3876 | .3867 | .3857 | .3847 | .3836 | .3825 |
| 0.3 | .3814 | .3802 | .3790 | .3778 | .3765 | .3752 | .3739 | .3725 | .3712 | .3697 |
| 0.4 | .3683 | .3668 | .3653 | .3637 | .3621 | .3605 | .3589 | .3572 | .3555 | .3538 |
| 0.5 | .3521 | .3503 | .3485 | .3467 | .3448 | .3429 | .3410 | .3391 | .3372 | .3352 |
| 0.6 | .3332 | .3312 | .3292 | .3271 | .3251 | .3230 | .3209 | .3187 | .3166 | .3144 |
| 0.7 | .3123 | .3101 | .3079 | .3056 | .3034 | .3011 | .2989 | .2966 | .2943 | .2920 |
| 0.8 | .2897 | .2874 | .2850 | .2827 | .2803 | .2780 | .2756 | .2732 | .2709 | .2685 |
| 0.9 | .2661 | .2637 | .2613 | .2589 | .2565 | .2541 | .2516 | .2492 | .2468 | .2444 |
| 1.0 | .2420 | .2396 | .2371 | .2347 | .2323 | .2299 | .2275 | .2251 | .2227 | .2203 |
| 1.1 | .2179 | .2155 | .2131 | .2107 | .2083 | .2059 | .2036 | .2012 | .1989 | .1965 |
| 1.2 | .1942 | .1919 | .1895 | .1872 | .1849 | .1826 | .1804 | .1781 | .1758 | .1736 |
| 1.3 | .1714 | .1691 | .1669 | .1647 | .1626 | .1604 | .1582 | .1561 | .1539 | .1518 |
| 1.4 | .1497 | .1476 | .1456 | .1435 | .1415 | .1394 | .1374 | .1354 | .1334 | .1315 |
| 1.5 | .1295 | .1276 | .1257 | .1238 | .1219 | .1200 | .1182 | .1163 | .1145 | .1127 |
| 1.6 | .1109 | .1092 | .1074 | .1057 | .1040 | .1023 | .1006 | .0989 | .0973 | .0957 |
| 1.7 | .0940 | .0925 | .0909 | .0893 | .0878 | .0863 | .0848 | .0833 | .0818 | .0804 |
| 1.8 | .0790 | .0775 | .0761 | .0748 | .0734 | .0721 | .0707 | .0694 | .0681 | .0669 |
| 1.9 | .0656 | .0644 | .0632 | .0620 | .0608 | .0596 | .0584 | .0573 | .0562 | .0551 |
| 2.0 | .0540 | .0529 | .0519 | .0508 | .0498 | .0488 | .0478 | .0468 | .0459 | .0449 |
| 2.1 | .0440 | .0431 | .0422 | .0413 | .0404 | .0396 | .0387 | .0379 | .0371 | .0363 |
| 2.2 | .0355 | .0347 | .0339 | .0332 | .0325 | .0317 | .0310 | .0303 | .0297 | .0290 |
| 2.3 | .0283 | .0277 | .0270 | .0264 | .0258 | .0252 | .0246 | .0241 | .0235 | .0229 |
| 2.4 | .0224 | .0219 | .0213 | .0208 | .0203 | .0198 | .0194 | .0189 | .0184 | .0180 |
| 2.5 | .0175 | .0171 | .0167 | .0163 | .0158 | .0154 | .0151 | .0147 | .0143 | .0139 |
| 2.6 | .0136 | .0132 | .0129 | .0126 | .0122 | .0119 | .0116 | .0113 | .0110 | .0107 |
| 2.7 | .0104 | .0101 | .0099 | .0096 | .0093 | .0091 | .0088 | .0086 | .0084 | .0081 |
| 2.8 | .0079 | .0077 | .0075 | .0073 | .0071 | .0069 | .0067 | .0065 | .0063 | .0061 |
| 2.9 | .0060 | .0058 | .0056 | .0055 | .0053 | .0051 | .0050 | .0048 | .0047 | .0046 |
| 3.0 | .0044 | | | | | | | | | |
| 4.0 | .0001 | | | | | | | | | |

*Solution.* The area table is based on the *standard* normal curve, with mean of 0, height of mean, .3989, and standard deviation of 1. The area required in the example is between (0–7) and (10–7) each side of the mean or between $-7$ and $+3$. Further, the standard deviation in the example is 4.

Hence, the area required is between $-7/4$ of the standard deviation and 3/4 of the standard deviation. That is, between $-1.75$ and .75 of the *standard* normal curve. From Table 7.1, the area from 0 to 1.75 is .4599, and the area from 0 to .75 is .2734, making .7333 in total.

Now, in the example, the height of the mean is 4.9, compared with .3989 in the standard normal curve, and the standard deviation is 4, compared with 1 in the standard normal curve. Hence the required area is

$$.7333 \times \frac{4.9}{.3989} \times \frac{4}{1} = 36.0$$

EXAMPLE 7.11. A normal curve has mean of 10 and standard deviation of 4. The height of the mean is 8. What is the height of the curve at the following points on the $x$ axis: 0, 3, 6, 9, 12, and 15?

*Solution.* The values of $x$, calculated from the mean of 10 are

$$x = -10, \quad -7, \quad -4, \quad -1, \quad 2, \quad 5$$

$$z = \frac{x}{\sigma} = \frac{-10}{4}, \quad \frac{-7}{4}, \quad \frac{-4}{4}, \quad \frac{-1}{4}, \quad \frac{2}{4}, \quad \frac{5}{4}$$

$$= -2.5, \quad -1.75, \quad -1, \quad -.25, \quad .5, \quad 1.25$$

The height of the standard normal curve at these points are, from Table 7.2,

.0175, .0863, .2420, .3867, .3521, .1826

Since the height of the mean in the example is 8, compared to .3989 in the normal curve, these heights must be multiplied by 8/.3989 or 20.06, giving

| $x$ | 0 | 3 | 6 | 9 | 12 | 15 |
|---|---|---|---|---|---|---|
| **Height** | 0.35 | 1.73 | 4.85 | 7.76 | 7.06 | 3.66 |

**7.9 Other Theoretical Distributions.** There are other theoretical distributions, such as the multinomial and the negative binomial. However, the three distributions discussed in this chapter are by far the most important.

**7.10. Fitting Theoretical Distributions.** It is most useful to be able to substitute a theoretical model for an empirical distribution. This is usually done by determining (from study of the empirical distribution, from theoretical consideration, or both) the most appropriate model, and then assuming that the mean standard deviation of the empirical distribution are the mean and standard deviation of the model. This is discussed further in Chapter 8.

## Problems

**Problem 7.1.** If the chance of it raining on any day at a certain time of year is 1/3 what are the probabilities of 0, 1, 2, and 3 wet days in a period of three days? Assume that the weather on any day is independent of the weather on any other day, although, in fact, there is usually considerable correlation between the weather on successive days.

**Problem 7.2.** What are the mean and the standard deviation of the distribution in Problem 7.1?

**Problem 7.3.** What is the frequency distribution of the score, when two dice are thrown a large number of times?

**Problem 7.4.** Two dice are thrown ten times, and the number of times the total on the two dice is 4 is recorded. What is the frequency distribution of the number of times this total is recorded, if the experiment is carried out a large number of times? (Express your answer in a formula.)

**Problem 7.5.** What are the probabilities of 0, 1, and 2 fours, in the previous problem?

**Problem 7.6.** A pack of eleven cards, numbered 1 to 11 inclusive, is well shuffled; a card is drawn and its number noted. The card is returned to the pack and the procedure repeated. After 8 such drawings, a total of 0, 1, 2, . . . 7 or 8 threes may have been drawn. What is the frequency distribution of the number of threes drawn?

**Problem 7.7.** What is the mean and the standard deviation of the distribution in Problem 7.6?

**Problem 7.8.** A real estate broker sells an average of 3 houses a week. Assuming a Poisson distribution, what will be the probability that he will sell exactly 9 houses in 3 weeks? (The answer may be left in formula form.)

**Problem 7.9.** Calculate the Poisson distribution frequencies for $x = 0, 1, 2, 3, 5, 10$, and 15 for $\lambda = 1$ and for $\lambda = 10$.

**Problem 7.10.** Among a group of similar automobile drivers, the annual rate of accident involvement is 15%. For each driver in the group, the probability of being involved in an accident on any day is the same. If the accident experience follows the Poisson distribution, what is the proportion of drivers who are accident free, and what is the proportion involved in exactly two accidents in a year?

**Problem 7.11.** Calculate the properties of the binomial distribution for $p = 1/24$ and $n = 96$ and compare them with the properties of the Poisson distribution for $\lambda = 4$.

**Problem 7.12.** For what values of $\lambda$ in a Poisson distribution is the frequency at $x = 0$ greater than the frequency at any other value?

**Problem 7.13.** What are the differences between the binomial and the normal distributions?

**Problem 7.14.** What is the breadth of a central segment of the normal distribution which includes (1) 50%, and (2) 75% of the total frequency?

**Problem 7.15.** What is the height of the ordinate corresponding to the outer edges of the 50% and 75% zones, referred to in Problem 7.14, if the height of the mean is 1?

**Problem 7.16.** Is the normal distribution leptokurtic or platykurtic?

**Problem 7.17.** Is the binomial distribution leptokurtic or platykurtic?

**Problem 7.18.** Is the Poisson distribution leptokurtic or platykurtic?

**Problem 7.19.** Find the area of the standard normal curve between $z = -.67$ and $z = +.33$.

**Problem 7.20.** A normal distribution has a mean of $x = 9.6$ and a standard deviation of 2.1. The height of the mean is $y = 11.5$. What is the area under the curve between $x = 9$ and $x = 10$?

## Solutions

**Problem 7.1.** In Formula 7.1, $p = 1/3$, $q = 2/3$, and $n = 3$.

| x | x! | (n − x)! | n! | $\frac{n!}{x!(n-x)!}$ | $27(p^x q^{n-x})$ | 27 f(x) |
|---|---|---|---|---|---|---|
| 0 | 1 | 6 | 6 | 1 | 8 | 8 |
| 1 | 1 | 2 | 6 | 3 | 5 | 12 |
| 2 | 2 | 1 | 6 | 3 | 2 | 6 |
| 3 | 6 | 1 | 6 | 1 | 1 | 1 |
| | | | | | | 27 |

The probabilities are 8/27, 12/27, 6/27, and 1/27, respectively.

**Problem 7.2.**

| x | $x^2$ | f | fx | $fx^2$ |
|---|---|---|---|---|
| 0 | 0 | 8 | 0 | 0 |
| 1 | 1 | 12 | 12 | 12 |
| 2 | 4 | 6 | 12 | 24 |
| 3 | 9 | 1 | 3 | 9 |
| | | 27 | 27 | 45 |

$$\text{Mean} = \frac{27}{27} = 1$$

$$m_2' = \frac{45}{27} = 1\tfrac{2}{3}$$

$$\text{Variance} = 1\tfrac{2}{3} - 1^2 = \tfrac{2}{3}$$

$$\text{Standard Deviation} = \sqrt{\frac{2}{3}} = .82$$

**Problem 7.3.** From the methods of Chapter 6, the relative frequencies of various scores are:

| **Scores** | 2 | 3 | 4 | 5 | 6 | 7 | 8 | 9 | 10 | 11 | 12 |
|---|---|---|---|---|---|---|---|---|---|---|---|
| **Relative Frequency** | 1 | 2 | 3 | 4 | 5 | 6 | 5 | 4 | 3 | 2 | 1 |

corresponding to the number of ways in which each score can occur.

The probabilities are these relative frequencies divided by the total frequency of 36. It should be noted that the distribution, while based on probabilities, is *not* a binomial distribution.

**Problem 7.4.** The probability of getting a 4 at a single throw of two dice is (from the answer to Problem 7.3)

$$\frac{3}{36} \text{ or } \frac{1}{12}$$

From Formula 7.1, the frequency distribution of 4 in ten throws is

$$f(x) = \frac{10!}{x!\,(10 - x)!}\left(\frac{1}{12}\right)^x\left(\frac{11}{12}\right)^{10-x}$$

*This distribution is a binomial distribution.*

**Problem 7.5.** We shall need to evaluate $(1/12)^x(11/12)^{10-x}$ for $x = 0$, 1, and 2. This is best done by logarithms; log 11 = 1.04139, log 12 = 1.07918.

| **x** (1) | **(10 – x)** (2) | **(10 – x) log 11** (3) | **10 log 12** (4) | **(3) – (4)** (5) | **antilog (5)** (6) |
|---|---|---|---|---|---|
| 0 | 10 | 10.4139 | 10.7918 | 9.6221-10 | .4189 |
| 1 | 9 | 9.3725 | 10.7918 | 8.5807-10 | .03808 |
| 2 | 8 | 8.3311 | 10.7918 | 7.5393-10 | .00346 |

The calculation then proceeds as follows:

| **x** | **x!(10 – x)!** | $\frac{10!}{x!(10-x)!}$ | $\left(\frac{1}{12}\right)^x\left(\frac{11}{12}\right)^{10-x}$ | **f(x)** |
|---|---|---|---|---|
| 0 | 10! | 1 | .4189 | .4189 |
| 1 | 9! | 10 | .03808 | .3808 |
| 2 | 2 × 8! | 45 | .00346 | .1557 |

The required probabilities are .42, .38, and .16.

**Problem 7.6.** From Formula 7.1, the distribution is

$$\frac{10^8}{11^8},\quad \frac{8!}{7!1!} \times \frac{10^7}{11^8},\quad \frac{8!}{6!2!} \times \frac{10^6}{11^8}, \ldots$$

giving a frequency distribution of

| **No. of threes** | **Relative Frequency** |
|---|---|
| 0 | $10^8$ |
| 1 | $8 \times 10^7$ |
| 2 | $28 \times 10^6$ |
| 3 | $56 \times 10^5$ |
| 4 | $70 \times 10^4$ |
| 5 | $56 \times 10^3$ |

| No. of threes | Relative Frequency |
|---|---|
| 6 | $28 \times 10^2$ |
| 7 | $8 \times 10$ |
| 8 | 1 |

The probabilities are $\frac{1}{11^8}$ times these figures and add to unity.

**Problem 7.7.** The distribution is binomial with

$$p = \frac{1}{11}, q = \frac{10}{11}, \text{ and } n = 8$$

$$\text{Mean} = 8 \times \frac{1}{11} = .73$$

$$\text{Variance} = 8 \times \frac{1}{11} \times \frac{10}{11} = .66$$

$$\text{Standard Deviation} = \sqrt{.66} = .81$$

**Problem 7.8.** Since the average rate of sale is 3 houses a week, the average for 3 weeks is 9. Hence, the mean of the distribution for a 3 week period = 9.

The probability of selling exactly 9 houses in 9 weeks is

$$f(9) = \frac{9^9 e^{-9}}{9!}$$

$$= .13$$

**Problem 7.9.** First calculate the values of $e^{-\lambda}$

$$\log e = .043429 \qquad \log e = 0.43429$$

$$1 \times \log e = 0.43429 \qquad 10 \times \log e = 4.3429$$

$$-1 \times \log e = 9.56571\text{-}10 \qquad -10 \times \log e = 5.6571\text{-}10$$

$$e^{-1} = .3679 \qquad e^{-10} = 4.54 \times 10^{-5}$$

For $\lambda = 1$,

| x | $\lambda^x$ | $e^{-\lambda}$ | x! | f |
|---|---|---|---|---|
| 0 | 1 | .3679 | 1 | .3679 |
| 1 | 1 | .3679 | 1 | .3679 |
| 2 | 1 | .3679 | 2 | .1839 |
| 3 | 1 | .3679 | 6 | .0613 |
| 5 | 1 | .3679 | 120 | .0031 |
| 10 | 1 | .3679 | $3.6 \times 10^6$ | .0000 |
| 15 | 1 | .3679 | $1.3 \times 10^{12}$ | .0000 |

For $\lambda = 10$,

| x | $\lambda^x$ | $e^{-\lambda}$ | x! | f |
|---|---|---|---|---|
| 0 | 1 | $4.54 \times 10^{-5}$ | 1 | .0000 |
| 1 | 10 | $4.54 \times 10^{-5}$ | 1 | .0005 |
| 2 | 100 | $4.54 \times 10^{-5}$ | 2 | .0023 |
| 3 | 1000 | $4.54 \times 10^{-5}$ | 6 | .0076 |
| 5 | $10^5$ | $4.54 \times 10^{-5}$ | 120 | .0378 |
| 10 | $10^{10}$ | $4.54 \times 10^{-5}$ | $3.63 \times 10^6$ | .1251 |
| 15 | $10^{15}$ | $4.54 \times 10^{-5}$ | $1.31 \times 10^{12}$ | .0347 |

**Problem 7.10.** This is a Poisson distribution with mean of .15. Therefore, $\lambda = .15$. The first three terms of the distribution are calculated from Formula 7.2.

$$\begin{aligned} \log e &= 0.43429 \\ .15 \log e &= 0.06514 \\ -.15 \log e &= 9.93486\text{-}10 \\ e^{-.15} &= .861 \end{aligned}$$

| x | $\lambda^x$ | $e^{-.15}$ | x! | f |
|---|---|---|---|---|
| 0 | 1 | .861 | 1 | .861 |
| 1 | .15 | .861 | 1 | .129 |
| 2 | .0225 | .861 | 2 | .010 |

86% of the drivers will be accident free and one in a hundred will be involved in two accidents during the period of a year.

**Problem 7.11.**

| | **Binomial** | **Poisson** |
|---|---|---|
| Mean | 4 | 4 |
| Variance | 3.83 | 4 |
| Standard Deviation | 1.96 | 2 |
| Moment coefficient of skewness | 4.7 | 5 |
| Moment coefficient of kurtosis | 3.2 | 3.25 |

It will be noted how close the two distributions become when $p$ is small.

**Problem 7.12.** If tables of the Poisson distribution are available, it is seen that for $\lambda = 1$, $f(0) = f(1)$, the required answer is for all values of $\lambda$ *less than* 1. If tables are not available, it is necessary to calculate the value of $\lambda$ corresponding to $f(0) = f(1)$. Now

$$f(0) = \frac{\lambda^0 c^{-\lambda}}{0!} = e^{-\lambda}$$

and $$f(1) = \frac{\lambda c^{-\lambda}}{1!} = \lambda c^{-\lambda}$$

giving $\lambda = 1$ and $f(0)$ will be greater than $f(1)$ for all values of $\lambda$ less than 1.

**Problem 7.13.** The essential difference between the binomial and the normal distributions is that the former is discrete and not symmetrical except when $p = q = .5$, while the latter is continuous and symmetrical. However, when $n$ is large, the normal distribution provides a close approximation to the binomial distribution, unless $p$ or $q$ are small.

**Problem 7.14.** From Table 7.1, it is seen that the ordinate $z$ corresponding to an area of .25 between $0$–$z$ of the normal curve is .674. Hence, the breadth of the central section including 50% of the frequencies is $2 \times .674$ or $1.35\sigma$. For 75%, the breadth is $2 \times 1.15$ or $2.3\sigma$.

**Problem 7.15.**

| Edge of Zone (1) | z (2) | Ordinates from Table 7.2 (3) | Col. (3) ÷ .3989 |
|---|---|---|---|
| 50% | .674 | .318 | .80 |
| 75% | 1.15 | .206 | .52 |

It is necessary to divide by .3989 since this figure, and not 1, is the height of the mean in Table 7.2.

**Problem 7.16.** The moment coefficient of kurtosis of the normal curve is 3, and by definition, the curve is neither leptokurtic or platykurtic. It is *mesokurtic*. (See Chapter 5.)

**Problem 7.17.** For values of $p$ between .79 and .21, the binomial distribution is platykurtic, and for values outside that range, the distribution is leptokurtic.

**Problem 7.18.** The Poisson distribution is always leptokurtic.

**Problem 7.19.** $.2486 + .1293 = .3779$

**Problem 7.20.**

$$(.1126 + .0753) \times 2.1 \times \frac{11.5}{.39894} = 11.4.$$

The value, .1126 is obtained by interpolation from Table 7.1.

# 8

# Curve Fitting

**8.1. Introduction.** It is often necessary to fit a smooth curve to empirical statistical data. The data may be in a form which can be represented fairly closely by one of the probability distributions considered in Chapter 7, or it may be in the form of a continuously increasing or decreasing function. While in many cases these latter functions will be time series, such as the cost of living on December 31st of successive years, this will not always be so. The distance in feet required to stop an automobile by the application of its brakes, plotted against the speed of the automobile when the brakes are first applied, is typical of the kind of increasing or decreasing function arising in statistical work, which is not a time series.

There are two standard methods of curve fitting.* The *graphic method*, where the smooth curve is drawn by hand, and the *formula method*, where an appropriate theoretical curve is assumed and mathematical procedures are used to provide maximum closeness of fit. When fitting a formula to a frequency distribution type of curve, it is usual to *equate the parameters* of the assumed distribution to those of the actual data. For an increasing or decreasing curve, the *method of least squares* is normally employed. Both these methods are explained later in the chapter.

**8.2. Theoretical Formulas for Increasing and Decreasing Functions.** Just as there are various theoretical probability distributions which can be used as models for empirical probability distri-

---

*There are other methods of *smoothing* data, see, for example, the moving average method in Chapter 9.

butions, so there are theoretical curves which can be used as models for empirical time series and other similar curves. By far the most usual model is the *straight line* which can be expressed by the following formula

(1) Linear $y = a_0 + a_1 x$

Other lines which may be used are

(2) Parabolic or 2nd order $y = a_0 + a_1 x + a_2 x^2$

(3) Cubic or higher order $y = a_0 + a_1 x + a_2 x^2 + a_3 x^3 + \cdots$

(4) Hyperbolic $\dfrac{1}{y} = a_0 + a_1 x$

(5) Exponential $\log y = a_0 + a_1 x$

(6) Geometric $y = ax^b$ or $a + bx^c$

EXAMPLE 8.1. Give an example of each of the formulas, (1), (2), (4), and (5) above, passing through the points $x = 0$, $y = 1$ and $x = 10$, $y = 2$. Tabulate the values of $y$ for $x = 0, 1, 2$, etc., ... 10, and for $x = 20$.

*Solution.* Formulas (1), (4), and (5) have only two unknowns, $a$ and $b$, so that, in each case, only one line can be drawn through the two points. Formula (2) has three unknowns, and more than one line can be drawn. For Formula (2), the line selected will be that produced by putting $a_1 = 0$, so that the formula becomes

$$y = a_0 + a_2 x^2$$

For the straight line, by substituting the given values $x = 0$, $y = 1$ and $x = 10$, $y = 2$ in Formula (1) we get

$$1 = a_0$$
$$2 = a_0 + 10a_1$$

Solving,

$$a_0 = 1, \text{ and } a_1 = \frac{1}{10}.$$

Proceeding in the same manner for the other formulas, the following equations are obtained:

(1) Straight line $y = 1 + \dfrac{1}{10} x$

(2) Parabola $y = 1 + \dfrac{1}{100} x^2$

(4) Hyperbola $\frac{1}{y} = 1 - \frac{1}{20}x$

(5) Exponential $\log y = (.03010)x$

The actual values are set out below.

| x | Straight Line (1) | Parabola (2) | Hyperbola (4) | Exponential (5) |
|---|---|---|---|---|
| 0 | 1.00 | 1.00 | 1.00 | 1.00 |
| 1 | 1.10 | 1.01 | 1.05 | 1.07 |
| 2 | 1.20 | 1.04 | 1.11 | 1.15 |
| 3 | 1.30 | 1.09 | 1.18 | 1.23 |
| 4 | 1.40 | 1.16 | 1.25 | 1.32 |
| 5 | 1.50 | 1.25 | 1.33 | 1.41 |
| 6 | 1.60 | 1.36 | 1.43 | 1.52 |
| 7 | 1.70 | 1.49 | 1.54 | 1.62 |
| 8 | 1.80 | 1.64 | 1.67 | 1.74 |
| 9 | 1.90 | 1.81 | 1.82 | 1.87 |
| 10 | 2.00 | 2.00 | 2.00 | 2.00 |
| ⋮ | | | | |
| 20 | 3.00 | 5.00 | Infinity | 4.00 |

This example shows clearly how very different the results will be according to the formula chosen. Obviously with only two values given, we have no indication of the true shape of the curve, and curve fitting is not really practical.

EXAMPLE 8.2. In the four curves given above, the value corresponding to $x = 5$ is 1.5 or less. Can an example be constructed where $x$ has a higher value?

*Solution.* In the previous example, we arbitrarily made $a_1 = 0$ for the parabola. If $a_1$ is given a positive value, the value of $a_2$ will become smaller, and when $a_1 = 1/10$, $a_2$ will equal 0. Increasing $a_1$ further, $a_2$ becomes negative, and the value of the function for $x = 5$ is greater than 1.5.

Thus, if the formula is

$$y = 1 + \frac{1}{5}x - \frac{1}{100}x^2$$

the values of the function are

| x | y |
|---|---|
| 0 | 1.00 |
| 5 | 1.75 |
| 10 | 2.00 |
| 20 | 1.00 |

It may be noted that, in this case, $y$ attains a maximum value at $x = 10$, and thereafter decreases.

**8.3. Graphic Method.** In using the graphic method to fit a smooth curve to empirical statistical data, the data must first be plotted on a large suitable piece of paper. No attempt should be made to make the smooth curve actually pass through the points tabulated; rather, the curve should pass between the points leaving, approximately, an equal number of points on either side of it. The curve should pass as close to the data as possible while remaining smooth. If the data lies approximately on a straight line, a rule may be used, and for other data, a flexible ruler or tracings of suitable smooth curves, if available, are helpful.

EXAMPLE 8.3. The following table gives the thousands of short tons of iron and steel exported by the U.S.A. from 1954 to 1966. Fit a smooth curve to the data by the graphic method.

| Year | Short Tons (000) | Year | Short Tons (000) | Year | Short Tons (000) |
|---|---|---|---|---|---|
| 1954 | 4,826 | 1959 | 7,009 | 1964 | 12,051 |
| 1955 | 9,723 | 1960 | 11,493 | 1965 | 8,934 |
| 1956 | 11,574 | 1961 | 12,442 | 1966 | 7,776 |
| 1957 | 13,657 | 1962 | 7,597 | | |
| 1958 | 6,328 | 1963 | 9,208 | | |

*Solution.* The data are plotted in Figure 8.1, and the smooth curve is drawn passing between the values.

EXAMPLE 8.4. The following table gives the observed frequencies of accidents in a four month period in a certain factory.

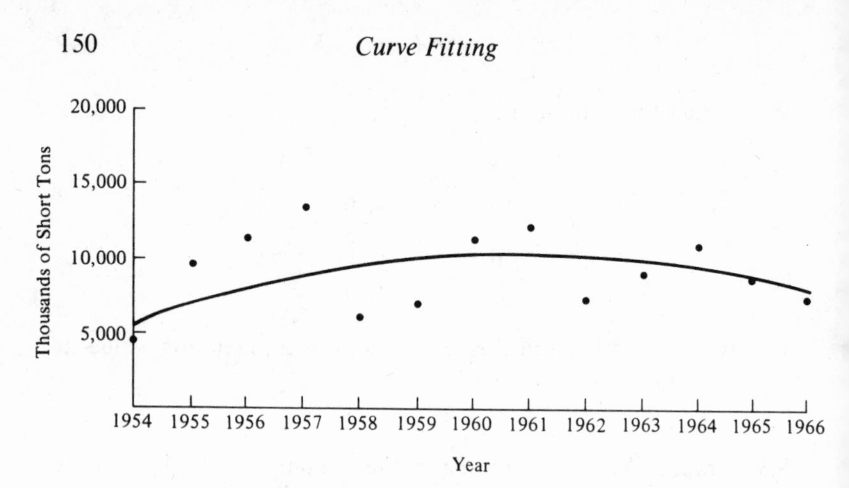


Figure 8.1. Steel exports. Graphic method of curve fitting.

| No. of Accidents | Observed Frequency | No. of Accidents | Observed Frequency |
|---|---|---|---|
| 0 | 239 | 8 | – |
| 1 | 98 | 9 | 4 |
| 2 | 57 | 10 | 1 |
| 3 | 33 | 11 | – |
| 4 | 9 | 12 | – |
| 5 | 2 | 13 | 1 |
| 6 | 2 | | 447 |
| 7 | 1 | | |

Fit a smooth curve to these data for accident frequency *3 or greater*.

*Solution.* The observed and smoothed frequencies are shown in Figure 8.2.

**8.4. Equating Parameters.** Except in the case of a straight line (when a ruler is used) the graphic method will not produce a curve which is perfectly smooth, and will only provide an approximation to one of the theoretical probability distributions. There are many advantages in fitting a theoretical distribution (rather than a similar hand-drawn smooth curve) to a given body of data. With a theoretical distribution, areas under portions of the curve, and various parameters, can be obtained from tables or readily calculated.

When it is desired to fit a normal, Poisson, or other distribution to given data, the mean and standard deviation of the ob-

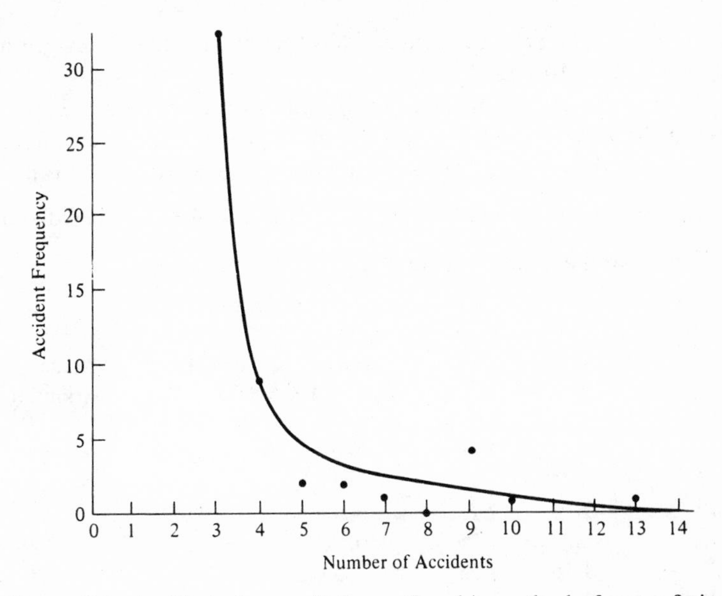


Figure 8.2. Accidents in soap factory. Graphic method of curve fitting.

served data are calculated, and then are used to determine the fitted formula.

EXAMPLE 8.5. Fit a normal curve to the following distribution of persons in a study of the mortality experience under life insurance policies.

| Age Group | Number in Thousands | Age Group | Number in Thousands |
|---|---|---|---|
| 0– | 51 | 40– | 1,946 |
| 5– | 38 | 45– | 1,229 |
| 10– | 53 | 50– | 641 |
| 15– | 172 | 55– | 290 |
| 20– | 667 | 60– | 102 |
| 25– | 1,286 | 65– | 28 |
| 30– | 1,996 | 70– | 6 |
| 35– | 2,319 | TOTAL | 10,824 |

*Solution.* Using $37\frac{1}{2}$ as the origin, and 5 years as the class interval, the mean is found to be 100/10,824, and the variance, 42,210/10,824. These figures for the mean and variance are very close to 0 and 4, respectively, so that for curve fitting we may assume that age $37\frac{1}{2}$ is the mean, and that the standard deviation is $5 \times \sqrt{4} = 10$ years.

Since the data are in age groups, the proper comparison between the observed figures and the fitted curve is the area under sections of the curve. The age group 35– corresponds to the range $-.25$ to $+.25$ for $\frac{x}{\sigma}$, and this and other areas are obtained from Table 7.1 and are set out below.

| Age Group | Normal Curve $\frac{x}{\sigma}$ | Normal Curve Area from Table 7.1 | Fitted Curve | Actual Distribution |
|---|---|---|---|---|
| 0– | $\infty$ to $-3.25$ | .0006 | 6 | 51 |
| 5– | $-3.25$ to $-2.75$ | .0024 | 26 | 38 |
| 10– | $-2.75$ to $-2.25$ | .0092 | 99 | 53 |
| 15– | $-2.25$ to $-1.75$ | .0279 | 302 | 172 |
| 20– | $-1.75$ to $-1.25$ | .0655 | 709 | 667 |
| 25– | $-1.25$ to $-.75$ | .1210 | 1,310 | 1,286 |
| 30– | $-.75$ to $-.25$ | .1747 | 1,891 | 1,996 |
| 35– | $-.25$ to $+.25$ | .1974 | 2,138 | 2,319 |
| 40– | $+.25$ to $+.75$ | .1747 | 1,891 | 1,946 |
| 45– | $+.75$ to $+1.25$ | .1210 | 1,310 | 1,229 |
| 50– | $+1.25$ to $+1.75$ | .0655 | 709 | 641 |
| 55– | $+1.75$ to $+2.25$ | .0279 | 302 | 290 |
| 60– | $+2.25$ to $+2.75$ | .0092 | 99 | 102 |
| 65– | $+2.75$ to $+3.25$ | .0024 | 26 | 28 |
| 70– | $+3.25$ to $\infty$ | .0006 | 6 | 6 |
| | | 1.0000 | 10,824 | 10,824 |

It will be seen that the normal curve provides a very satisfactory approximation to the actual distribution.

EXAMPLE 8.6. Attempt to fit a Poisson distribution to the full range of data (0–13) in Example 8.4.

*Solution.* By calculation, the mean of the data is .973 and the variance is 2.44. Since in the Poisson distribution these two parameters are the same, this immediately throws doubt on the likelihood of a satisfactory fit. Further for values of $\lambda$ greater than unity, the value corresponding to $x = 1$ is greater than the value corresponding to $x = 0$, while the reverse is the case with the data under consideration.

The following table shows three examples of attempting to fit a Poisson curve. In the first example, the value for $x = 0$ has been approximately equated, and in the second and third examples, the total frequencies have been equated. The latter is the more usual procedure. It will be seen that the *tail of the data* is too long to make the Poisson curve suitable.

| | | Trial Poisson Curves | | |
|---|---|---|---|---|
| **No. of Accidents** | **Observed Frequency** | **(1)** $\lambda = .5$ | **(2)** $\lambda = 1$ | **(3)** $\lambda = 1.5$ |
| 0 | 239 | 243 | 164 | 100 |
| 1 | 98 | 121 | 164 | 150 |
| 2 | 57 | 30 | 82 | 112 |
| 3 | 33 | 5 | 27 | 56 |
| ⋮ | ⋮ | ⋮ | ⋮ | ⋮ |
| 6 | 2 | 0 | 0 | 2 |
| ⋮ | ⋮ | ⋮ | ⋮ | ⋮ |
| 10 | 1 | 0 | 0 | 0 |
| **Total** | 447 | 400 | 447 | 447 |

**8.5. Least Squares Method.** If a definition of "*best fitting line*," or "*goodness of fit*" is established, formulas can be designed to determine this line, once its form (straight line, parabola, etc.) has been selected. The definition generally used in statistical work is the *least squares line*.

If the distances $D_1$, $D_2$, $D_3$, etc., parallel to the $y$ axis are measured from the empirical data points to the curve, then the least squares line is the one that makes $D_1^2 + D_2^2 + D_3^2 + \cdots$ a minimum.*

For a straight line,

$$y = a_0 + a_1 x$$

*The least squares line is not completely satisfactory because it gives too great a weight to extreme values. Some writers have proposed the least distance line, when $|D_1| + |D_2| + |D_3| + |D_4| + \cdots$ is a minimum. $|D|$ is the distance taken as positive in each case. In the past it has been difficult to calculate this line because of the sign problem, but this can be overcome with modern computers.

fitting $N$ points, $(X_1, Y_1)$, $(X_2, Y_2)$, etc., the sum of the squares of the distances is a minimum if

$$\begin{aligned} \Sigma Y &= a_0 N + a_1 \Sigma X \\ \Sigma XY &= a_0 \Sigma X + a_1 \Sigma X^2 \end{aligned} \tag{8.1}$$

These two equations are called the *normal equations of the least squares line.*

The corresponding equations for fitting a parabola,

$$y = a_0 + a_1 x + a_2 x^2$$

are

$$\begin{aligned} \Sigma Y &= a_0 N + a_1 \Sigma X + a_2 \Sigma X^2 \\ \Sigma XY &= a_0 \Sigma X + a_1 \Sigma X^2 + a_2 \Sigma X^3 \\ \Sigma X^2 Y &= a_0 \Sigma X^2 + a_1 \Sigma X^3 + a_2 \Sigma X^4 \end{aligned} \tag{8.2}$$

Solving Equations 8.1, the values of $a_0$ and $a_1$ are:

$$a_0 = \frac{(\Sigma Y)(\Sigma X^2) - (\Sigma X)(\Sigma XY)}{N(\Sigma X^2) - (\Sigma X)^2}$$

$$a_1 = \frac{N(\Sigma XY) - (\Sigma X)(\Sigma Y)}{N(\Sigma X^2) - (\Sigma X)^2}$$

EXAMPLE 8.7. Fit a straight line to the following data, for the U.S. Gross National Product, in billions of dollars.

| Year | G.N.P. | Year | G.N.P. |
|---|---|---|---|
| 1957 | 441.1 | 1962 | 560.3 |
| 1958 | 447.3 | 1963 | 590.5 |
| 1959 | 483.7 | 1964 | 632.4 |
| 1960 | 503.7 | 1965 | 683.9 |
| 1961 | 520.1 | 1966 | 743.3 |

*Solution.* Take 1961 as 0 in order to simplify the calculation, which can be set out in tabular form.

| Year | u | v | $u^2$ | uv | Fitted Curve $a_0 + a_1 u$ |
|---|---|---|---|---|---|
| 1957 | −4 | 441.1 | 16 | −1764.4 | 413.0 |
| 1958 | −3 | 447.3 | 9 | −1341.9 | 445.6 |
| 1959 | −2 | 483.7 | 4 | −967.4 | 478.5 |

| Year | u | v | $u^2$ | uv | Fitted Curve $a_0 + a_1 u$ |
|---|---|---|---|---|---|
| 1960 | −1 | 503.7 | 1 | −503.7 | 511.4 |
| 1961 | 0 | 520.1 | 0 | 0 | 544.2 |
| 1962 | 1 | 560.3 | 1 | 560.3 | 577.1 |
| 1963 | 2 | 590.5 | 4 | 1181.0 | 609.9 |
| 1964 | 3 | 632.4 | 9 | 1897.2 | 642.8 |
| 1965 | 4 | 683.9 | 16 | 2735.6 | 675.4 |
| 1966 | 5 | 743.3 | 25 | 3716.5 | 708.5 |
| **Totals** | 5 | 5606.3 | 85 | 5513.2 | 5606.3 |

The normal equations of the least squares are

$$5606.3 = 10\,a_0 + 5a_1$$
$$5513.2 = 5a_0 + 85a_1$$

Solving,

$$5606.3 = 10\,a_0 + 5a_1$$
$$11026.4 = 10\,a_0 + 170a_1$$

$$5420.1 = 165a_1$$
$$a_1 = 32.85$$
$$10a_0 = 5606.3 - 164.3 = 5442.0$$
$$a_0 = 544.2$$

EXAMPLE 8.8. The table below gives the U.S. wholesale price index for all farm products for the period 1955 to 1966. Find the least squares line for these data.

| Year | Index | Year | Index |
|---|---|---|---|
| 1955 | 97.9 | 1961 | 96.0 |
| 1956 | 96.6 | 1962 | 97.7 |
| 1957 | 99.2 | 1963 | 95.7 |
| 1958 | 103.6 | 1964 | 94.3 |
| 1959 | 97.2 | 1965 | 98.4 |
| 1960 | 96.9 | 1966 | 102.5 |

*Solution.* The data are so close to a level straight line just below 100, that a level line at the average figure of 98.0 is clearly appropriate. Applying the least squares method, and using a time origin of 1960 and an index origin of 100, the calculations give:

| Year | u | v | $u^2$ | uv | Horizontal Line | Least Squares Line |
|---|---|---|---|---|---|---|
| 1955 | −5 | −2.1 | 25 | +10.5 | −2.0 | −1.89 |
| 1956 | −4 | −3.4 | 16 | +13.6 | −2.0 | −1.91 |
| 1957 | −3 | −0.8 | 9 | +2.4 | −2.0 | −1.93 |
| 1958 | −2 | +3.6 | 4 | −7.2 | −2.0 | −1.95 |
| 1959 | −1 | −2.8 | 1 | +2.8 | −2.0 | −1.97 |
| 1960 | 0 | −3.1 | 0 | 0 | −2.0 | −1.99 |
| 1961 | 1 | −4.0 | 1 | −4.0 | −2.0 | −2.01 |
| 1962 | 2 | −2.3 | 4 | −4.6 | −2.0 | −2.03 |
| 1963 | 3 | −4.3 | 9 | −12.9 | −2.0 | −2.05 |
| 1964 | 4 | −5.7 | 16 | −22.8 | −2.0 | −2.07 |
| 1965 | 5 | −1.6 | 25 | −8.0 | −2.0 | −2.09 |
| 1966 | 6 | +2.5 | 36 | +15.0 | −2.0 | −2.11 |
| | 6 | −24.0 | 146 | −15.2 | −24.0 | −24.0 |

The normal equations of the least squares line are:

$$-24.0 = 12a_0 + 6a_1$$
$$-15.2 = 6a_0 + 146a_1$$

Solving,

$$-24.0 = 12a_0 + 6a_1$$
$$-30.4 = 12a_0 + 292a_1$$
$$-6.4 = 286a_1$$
$$a_1 = -.02$$
$$12a_0 = -24.0 + .12$$
$$a_0 = 1.99$$

The least squares line starts at 98.1 in 1955 and drops steadily to 97.9 in 1966.

**8.6. Advantages and Disadvantages of Graphic Method.** The principal advantages of the *graphic* or *free hand* method are

1. It is easy to use.
2. No calculations are required.
3. For a straight line, using a ruler, excellent results are quickly obtained.
4. Judgment and a knowledge of the shape of similar data can be brought into play.
5. It is often the only practical method.

The principal disadvantages are:

1. The curve is not completely smooth, except when a ruler is used.
2. The curve is subjective.
3. Two persons will obtain different results.

**8.7. Advantages and Disadvantages of Parameter Method.** The principal advantages of equating parameters as a means of curve fitting are:

1. The curve is completely smooth.
2. The method produces a curve which is well-known.
3. Tables of areas, ordinates, and various parameters are available for testing the fit and for deducing results concerning the data.
4. It is useful when bodies of similar data have to be compared.
5. There is room for judgment in the choice of the formula and in the selection of the parameters.

The principal disadvantages are:

1. The results are tied to a mathematical formula.
2. The method is more laborious than the graphic.
3. The method can be used only when a suitable curve is available, (however there are many more curves than the three discussed in Chapter 7).

**8.8. Advantages and Disadvantages of Least Squares Method.** The principal advantages of the least squares method are:

1. The curve is completely smooth.
2. It provides the "best fit."
3. A mathematical formula is easily interpreted.
4. The application of the method is independent of subjective considerations (except for the selection of a straight line or other mathematical formula).
5. The results are convenient for forecasting.

The principal disadvantages are:

1. The results are tied to a mathematical formula, which may not really represent the data.
2. There is no room for judgment.
3. Too much weight is given to unusual values.

**8.9. Nonlinear Data.** When time series and similar data are of a form which precludes the use of a straight line, other curves, such as a parabola, may be used. However, it is often found that by changing the vertical scale to (for example) a logarithm scale, a close fit by means of a straight line may be obtained. This can be done readily by use of special graph paper ruled for a logarithm scale.

EXAMPLE 8.9. The following table shows the accumulation of $1,000 at 5 percent interest compounded annually over a period of years. Draw a graph of these figures. Draw a second graph of the logarithm of the amounts, and fit a straight line to the latter results.

| Years | Amount | Years | Amount |
|---|---|---|---|
| 0 | $1,000 | 60 | $18,679 |
| 10 | 1,629 | 70 | 30,426 |
| 20 | 2,653 | 80 | 49,561 |
| 30 | 4,322 | 90 | 80,730 |
| 40 | 7,040 | 100 | 131,501 |
| 50 | 11,467 | | |

*Solution.* The graph of the original data is given in Figure 8.3. Expressing the amounts in logarithm form, we have:

| Years | Amount | Log Amount | Years | Amount | Log Amount |
|---|---|---|---|---|---|
| 0 | $1,000 | 3.0000 | 60 | 18,679 | 4.2714 |
| 10 | 1,629 | 3.2119 | 70 | 30,426 | 4.4832 |
| 20 | 2,653 | 3.4237 | 80 | 49,561 | 4.6951 |
| 30 | 4,322 | 3.6357 | 90 | 80,730 | 4.9070 |
| 40 | 7,040 | 3.8476 | 100 | 131,501 | 5.1189 |
| 50 | 11,467 | 4,0595 | | | |

It will be seen from the graph of the log amount against the number of years (as set out in Figure 8.4) that the data now lie on an exact straight line.

**8.10 Descriptive Terms for a Straight Line.** In view of the simplicity and the importance of the straight line as a means of representing time series and similar data, the following terms should be noted.

*Slope.* The slope of a straight line is the ratio of the vertical to the horizontal distance between two points on the line. (This

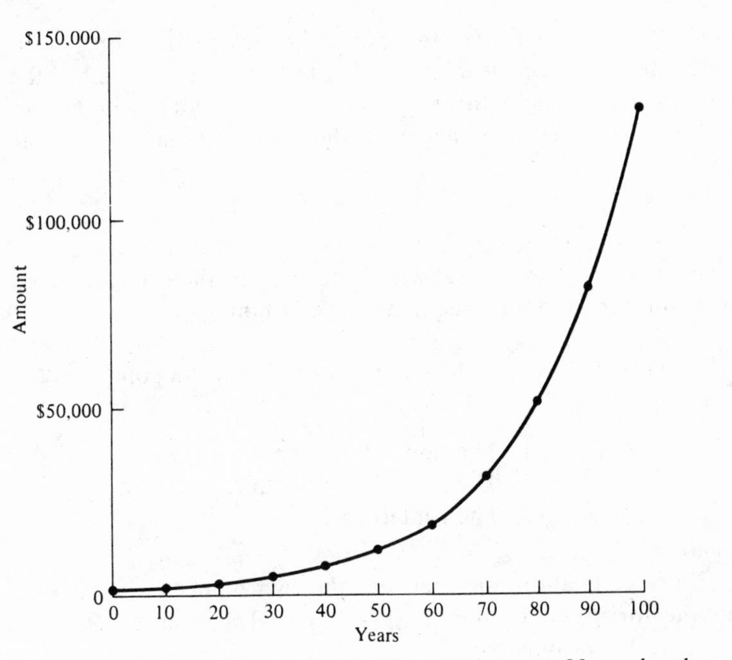


Figure 8.3. Accumulation of $1000 at 5% interest. Normal scale.

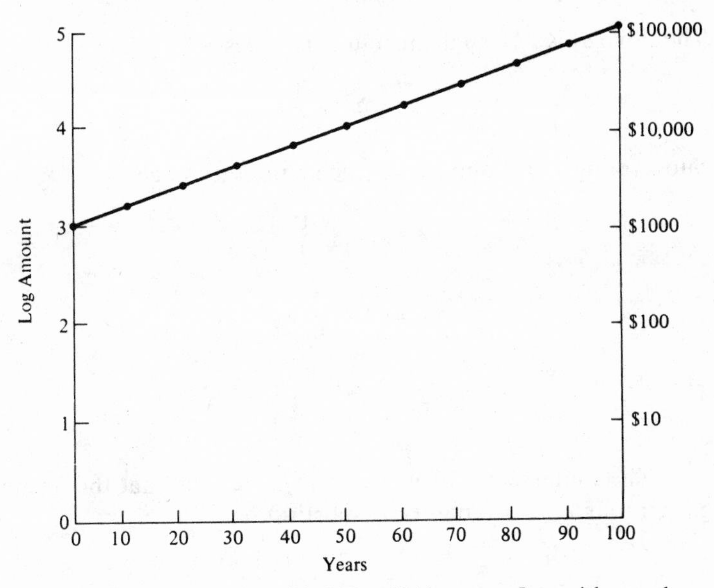


Figure 8.4. Accumulation of $1000 at 5% interest. Logarithm scale.

will be the same whatever two points are selected). For a horizontal line, the slope is 0; for a line at 45° to the horizontal, the slope will be 1; and for a vertical line, the slope will be $\infty$. When $y$ decreases as $x$ increases, the slope is negative. In the formula

$$y = a_0 + a_1 x$$

the constant $a_1$ is the slope.

*y intercept.* The point at which the line cuts the $y$ axis is called the $y$ intercept and this is equal to the constant $a_0$.

EXAMPLE 8.10. A straight line passes through the points $x = 1$, $y = 2$ and $x = 10$, $y = 3$. What is

1. The slope of the line,
2. The formula for the line,
3. The $y$ intercept?

*Solution.*

1. The vertical distance between the two points is $3 - 2 = 1$, and the horizontal distance is $10 - 1 = 9$. The slope is 1/9.

2. Let the formula be

$$y = a_0 + a_1 x$$

The slope $a_1$ is 1/9, so the formula becomes

$$y = a_0 + \frac{1}{9} x$$

Substituting $x = 1$ and $y = 2$ (the first point) we get

$$2 = a_0 + \frac{1}{9}$$

$$a_0 = 2 - \frac{1}{9} = \frac{17}{9}$$

The formula is

$$y = \frac{17}{9} + \frac{1}{8} x$$

3. The $y$ intercept is $a_0$ which is 17/9. Check that the formula goes through the two points by substitution.

$$2 = \frac{17}{9} + \frac{1}{9}$$

and

$$3 = \frac{17}{9} + \frac{10}{9}$$

## Problems

**Problem 8.1.** Two points completely define a straight line, or in other words, only one straight line can be drawn through two given points. How many points define

1. a parabola
2. a cubic
3. a hyperbola?

**Problem 8.2.** The slope of a straight line is 1/10 and the $y$ intercept is 1.

1. What is the formula for the line?
2. What is the value of $y$ corresponding to $x = 0, 5$, and 10?

**Problem 8.3.** Find the formula for a parabola passing through the points

$$(x = 0, y = 2)\ (x = 5, y = 3)\ (x = 10, y = 5)$$

**Problem 8.4.** Find the values of $a_1$, $a_2$ and $a_3$ in the formula

$$\frac{1}{y} = a_0 + a_1x + a_2x^2$$

passing through the three points in Problem 8.3.

**Problem 8.5.** The curves in Problem 8.3, and 8.4, both pass through the same three points. What are the values of $y$ corresponding to $x = -5$ and $x = 15$ for the two curves?

**Problem 8.6.** The following table from a certain study, gives the rate of daylight accident involvement per hundred million vehicle-miles for passenger automobiles, analyzed by age of car.

| Age of Car | Accident Involvement Rate | Age of Car | Accident Involvement Rate |
|---|---|---|---|
| 1–1.9 | 161 | 6–6.9 | 251 |
| 2–2.9 | 162 | 7–7.9 | 272 |
| 3–3.9 | 193 | 8–8.9 | 285 |
| 4–4.9 | 214 | 9–9.9 | 300 |
| 5–5.9 | 236 | 10 and over | 396 |

Make a graphic graduation of these data.

**Problem 8.7.** A pair of dice are thrown 50 times, and the score for each throw recorded.

| Score | Observed Frequency | Score | Observed Frequency |
|---|---|---|---|
| 2 | 1 | 8 | 7 |
| 3 | 2 | 9 | 6 |
| 4 | 3 | 10 | 3 |
| 5 | 6 | 11 | 2 |
| 6 | 9 | 12 | 0 |
| 7 | 11 | | |

Draw a smooth curve through these data.

**Problem 8.8.** Use the graphic method to fit a curve to the following data for the mortgage debt outstanding in the U.S.A. at December 31st, for the years shown. Amounts are in billions of dollars.

| Year | Mortgage Debt | Year | Mortgage Debt |
|---|---|---|---|
| 1940 | 36.5 | 1960 | 206.8 |
| 1945 | 35.6 | 1961 | 226.3 |
| 1950 | 72.8 | 1962 | 248.6 |
| 1955 | 129.9 | 1963 | 274.3 |
| 1956 | 144.5 | 1964 | 300.1 |
| 1957 | 156.5 | 1965 | 326.2 |
| 1958 | 171.8 | 1966 | 347.1 |
| 1959 | 190.8 | | |

**Problem 8.9.** The following table gives the number of telephone calls for each half-hour on a small office switchboard. Draw a smooth curve through these data.

| Time | No. of Calls | Time | No. of Calls |
|---|---|---|---|
| 8:30– | 10 | 12:30– | 35 |
| 9:00– | 30 | 1:00– | 22 |
| 9:30– | 46 | 1:30– | 30 |
| 10:00– | 56 | 2:00– | 38 |
| 10:30– | 54 | 2:30– | 60 |
| 11:00– | 63 | 3:00– | 57 |
| 11:30– | 52 | 3:30– | 60 |
| 12:00– | 42 | 4:00– | 58 |
| | | 4:30– | 32 |

Could a frequency distribution be fitted to these data?

**Problem 8.10.** A class of 30 students take a certain test and obtain the following scores out of a maximum of 100.

| Score | Number of Students | Score | Number of Students |
|---|---|---|---|
| 40 | 5 | 70 | 5 |
| 50 | 6 | 80 | 5 |
| 60 | 6 | 90 | 3 |

Draw a smooth curve through these data.

**Problem 8.11.** Students passing a certain examination are analysed according to the number of times they have previously sat for the examination (and failed). The results are as follows:

| No. of Failures Prior to Passing | No. of Students |
|---|---|
| 0 | 40 |
| 1 | 26 |
| 2 | 2 |
| 3 | 2 |
| 4 or more | 0 |

Fit a Poisson distribution to these data.

**Problem 8.12.** Fit a binomial distribution to the data in Problem 8.11.

**Problem 8.13.** Fit a normal curve to the data in Problem 8.7.

**Problem 8.14.** Use the method of least squares to fit a straight line to the following data.

| x | 0 | 1 | 2 | 3 | 4 | 5 |
|---|---|---|---|---|---|---|
| y | 1 | 2 | 4 | 3 | 5 | 4 |

**Problem 8.15.** The following table gives figures for the average sale per occupied room in hotels. Fit a least squares straight line to these data.

| Year | Sale | Year | Sale |
|---|---|---|---|
| 1955 | $7.50 | 1963 | $9.37 |
| | | 1964 | 9.53 |
| 1960 | 9.15 | 1965 | 9.71 |
| 1961 | 9.23 | 1966 | 10.03 |
| 1962 | 9.35 | | |

**Problem 8.16.** In the following data, $y$ increases continuously with $x$. Show that the least squares straight line obtained by

making $x$ the *dependent variable* instead of $y$ is not the same as the normal least squares straight line.

| x | 0 | 1 | 2 | 4 | 5 | 7 | 9 |
|---|---|---|---|---|---|---|---|
| y | 0 | 2 | 3 | 4 | 7 | 9 | 10 |

**Problem 8.17.** Why are the two lines obtained in Problem 8.16 different?

**Problem 8.18.** Fit a least squares straight line to the data in Problem 8.11.

**Problem 8.19.** The following table gives the amount of ordinary life insurance in force in the U.S.A. (in billions of dollars). Fit a straight line curve to these amounts, and compare the two results.

| Year | Amount ($ billions) | Year | Amount ($ billions) |
|---|---|---|---|
| 1900 | 6 | 1940 | 79 |
| 1910 | 12 | 1950 | 149 |
| 1920 | 32 | 1960 | 340 |
| 1930 | 78 | | |

**Problem 8.20.** Fit the following curve to the data in Problem 8.19, using the least squares method.

$$y = a_0 + a_1 x + a_2 x^2$$

## Solutions

**Problem 8.1.** The number of points required to define a curve is equal to the number of unknown constants, $a_0$, $a_1$, etc., in the equation for the curve. The number of points required to define the curves listed in the question are (1) 3; (3) 4; (3) 2. The last answer corresponds to the simple formula of a hyperbola given in Section 8.2. The general hyperbola requires 4 points to define it.

**Problem 8.2.**

1. $y = 1 + \frac{1}{10}x$

2. 1, 1½, and 2.

**Problem 8.3.** The formula for a parabola is

$$y = a_0 + a_1 x + a_2 x^2$$

Substituting the given values

$$2 = a_0$$
$$3 = a_0 + 5a_1 + 25a_2$$
$$5 = a_0 + 10a_1 + 100a_2$$

Hence,

$$1 = 5a_1 + 25a_2$$
$$3 = 10a_1 + 100a_2$$

Solving,

$$50a_2 = 1, a_2 = \frac{1}{50}$$
$$10a_1 = 1, a_1 = \frac{1}{10}.$$
$$y = 2 + \frac{1}{10}x + \frac{1}{50}x^2$$

**Problem 8.4.**

$$\frac{1}{2} = a_0$$
$$\frac{1}{3} = a_0 + 5a + 25a_2$$
$$\frac{1}{5} = a_0 + 10a_1 + 100a_2$$
$$\frac{1}{y} = \frac{1}{2} - \frac{11}{300}x + \frac{1}{1500}x^2$$

**Problem 8.5.**

| | | **x = −5** | **x = 15** |
|---|---|---|---|
| **First formula** | $y = 2 + \frac{1}{10}x + \frac{1}{50}x^2$ | $y = 2$ | $y = 8.0$ |
| **Second formula** | $\frac{1}{y} = \frac{1}{2} - \frac{11}{300}x + \frac{1}{1500}x^2$ | $y = 1.43$ | $y = 10.0$ |

**Problem 8.6.** Note that the center of the age intervals are 1.45, 2.45, etc. The last value is best omitted from the curve fitting because there is no means of estimating the mean age. However, the curve, if projected, should at least reach the value of 396. The fitted curve is shown in Figure 8.5.

**Problem 8.7.** The plot of the data and the fitted curve are shown in Figure 8.6.

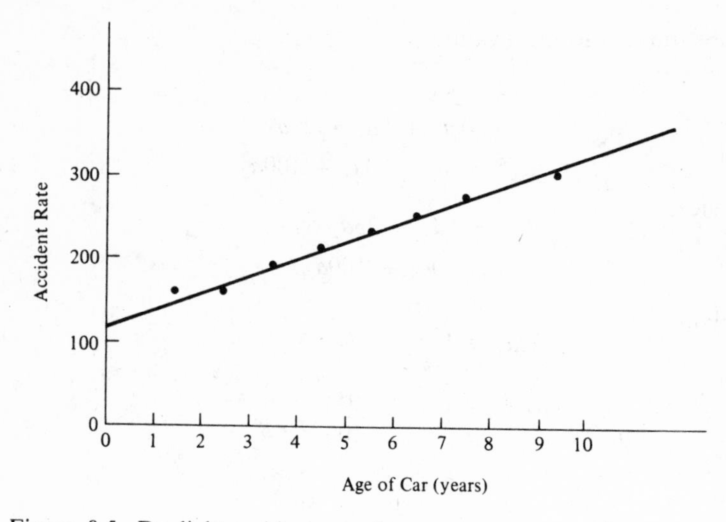


Figure 8.5. Daylight accident rates for passenger automobiles. Analysis by age of car.

**Problem 8.8.** It must be noted that data for the early years is given at quinquennial points only. These must, of course, be properly spaced out on the $x$ axis. Dates shown on the $x$ axis must indicate (by a footnote or other means) that they are

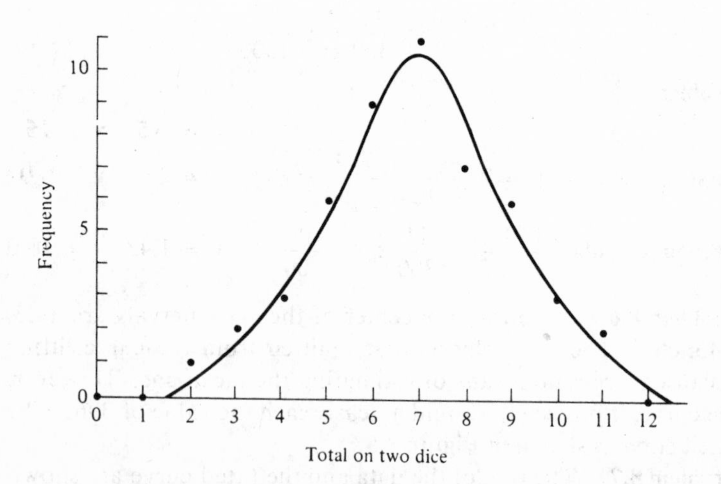


Figure 8.6. Distribution of scores from the throw of two dice.

December 31. This is best achieved as follows:

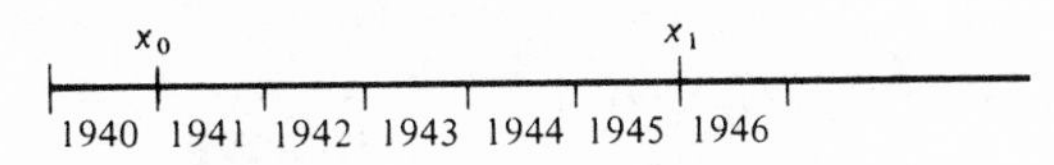


Figure 8.7. Method of indicating data at December 31st of any year.

The values for December 31, 1940 and December 31, 1945 are placed immediately above the points marked as $x_0$ and $x_1$.

**Problem 8.9.** Note that the figures given should be recorded against the midpoint of each interval: 8.45, 9.15, etc.

The distribution has two humps, one in the morning and one after lunch; for this reason none of the frequency distributions discussed in this book could be used. Two hump distribution formulas can however be constructed.

**Problem 8.10.** In plotting these data remember that the values for scores 30 and 100 are both zero. The smooth curve should not be greater than 1 or 2 at score 30 and must be zero for values above 100.

**Problem 8.11.** First calculate the mean and variance of the data.

| No. of Failures (x) | No. of Students (y) | xy | $x^2y$ |
|---|---|---|---|
| 0 | 40 | 0 | 0 |
| 1 | 26 | 26 | 26 |
| 2 | 2 | 4 | 8 |
| 3 | 2 | 6 | 18 |
| 4 | 0 | 0 | 0 |
| | 70 | 36 | 52 |

$$\text{Mean} = \frac{36}{70} = .51$$

$$\text{Variance} = \frac{52}{70} - (.51)^2$$

$$= .74 - .26 = .48$$

$\lambda = .5$ provides a close fit to both the mean and the variance. The values of the Poisson distribution for this value of $\lambda$ can be obtained from statistical tables or calculated.

| No. of Failures | No. of Students Observed | Fitted | Difference |
|---|---|---|---|
| 0 | 40 | 42.5 | +2.5 |
| 1 | 26 | 21.2 | −4.8 |
| 2 | 2 | 5.3 | +3.3 |
| 3 | 2 | 0.9 | −1.1 |
| 4 | 0 | 0.1 | +0.1 |
| | 70 | 70 | 0 |

**Problem 8.12.** The mean of the data is .51 and the variance is .48. From the formulas for the binomial curve,

$$\text{Mean} = np$$
$$\text{Variance} = npq$$

This gives

$$q = \frac{.48}{.51} = .94$$

Using this figure, we get $p = .06$ and $n = 8$. In view of the small volume of the data (70 observed cases), any attempt to fit the data too closely would be wrong and it will be sufficient to simplify the calculation by assuming $q = .9$, $p = .1$ and $n = 5$, giving

| x | x! | (n − x)! | n! | $\frac{n!}{x!(n-n)!}$ | $p^x$ | $q^x$ | Frequency | Fitted Curve | Observed |
|---|---|---|---|---|---|---|---|---|---|
| 0 | 1 | 120 | 120 | 1 | 1 | .59 | .59 | 41.3 | 40 |
| 1 | 1 | 24 | 120 | 5 | .1 | .66 | .33 | 23.1 | 26 |
| 2 | 2 | 6 | 120 | 10 | .01 | .73 | .07 | 4.9 | 2 |
| 3 | 6 | 2 | 120 | 10 | .001 | .81 | .01 | 0.7 | 2 |
| 4 | 24 | 1 | 120 | 5 | .0001 | .9 | — | 0 | 0 |
| 5 | 120 | 1 | 120 | 1 | .00001 | 1 | — | 0 | 0 |
| | | | | | | | 100 | 70 | 70 |

**Problem 8.13.** Proceed in the usual manner to calculate the mean and standard deviation of the data. (Use 7 as the origin). The mean is 6.86 and the

$$\text{standard deviation} = \frac{207}{50} - (.14)^2 = \sqrt{4.14 - .02}$$
$$= 2.03$$

Using mean of 7 and standard deviation of 2 we have

| **Point values required** | 7± | 0 | 1 | 2 | 3 | 4 | 5 |
|---|---|---|---|---|---|---|---|
| $\frac{x}{\sigma}$ | | 0 | .5 | 1.0 | 1.5 | 2.0 | 2.5 |
| **Ordinates from Table 7-2** | | .3989 | .3521 | .2420 | .1295 | .0540 | .0175 |

The negative values have been omitted because they are the same as the positive. The sum of the ordinates (positive and negative) is

$$.3989 + 2(.3521 + .2420 + .1295 + .0540 + .0175) = 1.9891$$

The ordinates from Table 7.2 must be increased in the ratio, $50 \div 1.9891 = 25$ approximately, to make the total agree with the curve to be fitted, giving

| | **Frequency** | | | **Frequency** | |
|---|---|---|---|---|---|
| **Score** | **Observed** | **Fitted** | **Score** | **Observed** | **Fitted** |
| 2 | 1 | 0.4 | 8 | 7 | 8.8 |
| 3 | 2 | 1.4 | 9 | 6 | 6.1 |
| 4 | 3 | 3.3 | 10 | 3 | 3.3 |
| 5 | 6 | 6.1 | 11 | 2 | 1.4 |
| 6 | 9 | 8.8 | 12 | 0 | 0.4 |
| 7 | 11 | 10.0 | | | |

**Problem 8.14.**

| x | y | $x^2$ | xy | **Least squares line** |
|---|---|---|---|---|
| 0 | 1 | 0 | 0 | 1.5 |
| 1 | 2 | 1 | 2 | 2.2 |
| 2 | 4 | 4 | 8 | 2.8 |
| 3 | 3 | 9 | 9 | 3.5 |
| 4 | 5 | 16 | 20 | 4.2 |
| 5 | 4 | 25 | 20 | 4.8 |
| 15 | 19 | 55 | 59 | |

Normal equations for the least square line are

$$19 = 6a_0 + 15a_1$$
$$59 = 15a_0 + 55a_1$$

Solving $\quad a_0 = 1.52 \qquad a_1 = .66$

The values are set out on the right hand side of the table above.

**Problem 8.15.** Use 1960 as origin and measure sales in difference from \$9.50, to reduce the arithmetic.

| Year | u | v | u² | uv | Least Squares Line From \$9.50 | Least Squares Line From Zero |
|---|---|---|---|---|---|---|
| 1955 | −5 | −2.00 | 25 | \$10.00 | −\$1.74 | \$7.76 |
| 1960 | 0 | − .35 | 0 | 0 | − .69 | 8.81 |
| 1961 | 1 | − .27 | 1 | − .27 | − .48 | 9.02 |
| 1962 | 2 | − .15 | 4 | − .30 | − .27 | 9.23 |
| 1963 | 3 | − .13 | 9 | − .39 | − .06 | 9.44 |
| 1964 | 4 | + .03 | 16 | + .12 | + .15 | 9.65 |
| 1965 | 5 | + .21 | 25 | +1.05 | + .36 | 9.86 |
| 1966 | 6 | + .53 | 36 | +3.18 | + .57 | 10.07 |
| | 16 | −\$2.13 | 116 | \$13.39 | −\$2.16 | |

$$-2.13 = 8a_0 + 16a_1$$

$$+13.39 = 16a_0 + 116a_1$$

$$a_0 = -.69 \text{ cents}, \qquad a_1 = .21 \text{ cents}$$

**Problem 8.16.** The equations for the two lines can be calculated in a single tabulation. The point $x = 5, y = 5$ is used as origin.

| (x − 5) | (y − 5) | (x − 5)² | (x − 5)(y − 5) | (y − 5)² |
|---|---|---|---|---|
| −5 | −5 | 25 | 25 | 25 |
| −4 | −3 | 16 | 12 | 9 |
| −3 | −2 | 9 | 6 | 4 |
| −1 | −1 | 1 | 1 | 1 |
| 0 | 2 | 0 | 0 | 4 |
| 2 | 4 | 4 | 8 | 16 |
| 4 | 5 | 16 | 20 | 25 |
| −7 | 0 | 71 | 72 | 84 |

For the least squares line, with $y$ as the dependent variable,

$$0 = 7a_0 - 7a_1$$

$$72 = -7a_0 + 71a_1$$

whence

$$a_0 = 1.125, \qquad a_1 = 1.125$$

giving

$$(y - 5) = 1.125 + 1.125(x - 5)$$

or

$$y = 0.5 + 1.125x$$

For the least squares line with $x$ as the dependent variable,

$$-7 = 7b_0 + 0$$
$$72 = 0 + 84b_1$$

whence

$$b_0 = -1, \qquad b_1 = .857$$

giving

$$(x - 5) = -1 + .857(y - 5)$$

or

$$x = -0.285 + .857y$$

which may be written

$$y = \frac{1}{.857}x + \frac{0.285}{.857} = .33 + 1.17x$$

**Problem 8.17.** The reason why the two equations obtained in Problem 8.16 are not identical is because the least squares line is obtained by minimizing the sum of the squares of the distances from the points to the line *measured parallel to the $y$ axis.* In the second set of calculations, the measurements are made parallel to the $x$ axis.

**Problem 8.18.** The least squares equation is

$$y = 37 - 13x$$

giving values

| **x** | 0 | 1 | 2 | 3 |
|---|---|---|---|---|
| **Observed** | 40 | 26 | 2 | 2 |
| **Fitted** | 37 | 24 | 11 | −2 |

This is not a satisfactory fit.

**Problem 8.19.** Use origin 1930 and 10 year intervals for $x$. The results are:

| **Year** | 1900 | 1910 | 1920 | 1930 | 1940 | 1950 | 1960 |
|---|---|---|---|---|---|---|---|
| **Actual Figures** | 6 | 12 | 32 | 78 | 79 | 149 | 340 |
| **Straight Line** | −41 | 6 | 53 | 100 | 147 | 194 | 241 |
| **Straight Line Log** | 7 | 13 | 26 | 49 | 93 | 178 | 339 |

**Problem 8.20.** The formulas for the normal equations are given in Section 8.5.
Using origin year 1930 and 10 year intervals,

| Year | u | $u^2$ | $u^3$ | $u^4$ | v | uv | $u^2v$ |
|---|---|---|---|---|---|---|---|
| 1900 | −3 | 9 | −27 | 81 | 6 | −18 | 54 |
| 1910 | −2 | 4 | −8 | 16 | 12 | −24 | 48 |
| 1920 | −1 | 1 | −1 | 1 | 32 | −32 | 32 |
| 1930 | 0 | 0 | 0 | 0 | 78 | 0 | 0 |
| 1940 | 1 | 1 | 1 | 1 | 79 | 79 | 79 |
| 1950 | 2 | 4 | 8 | 16 | 149 | 298 | 596 |
| 1960 | 3 | 9 | 27 | 81 | 340 | 1,020 | 3,060 |
| | 0 | 28 | 0 | 196 | 696 | 1,323 | 3,869 |

$$696 = 7a_0 + 28a_2$$
$$1{,}323 = 28a_1$$
$$3{,}845 = 28a_0 + 196a_2$$

Hence,

$$a_0 = 47.8$$
$$a_1 = 47.2$$
$$a_2 = 12.9$$

giving

| Year | 1900 | 1910 | 1920 | 1930 | 1940 | 1950 | 1960 |
|---|---|---|---|---|---|---|---|
| **Actual Figure** | 6 | 12 | 32 | 78 | 79 | 149 | 340 |
| **Parabola** | 22 | 5 | 14 | 48 | 108 | 194 | 306 |

It will be noted that this parabola is less satisfactory than fitting a straight line to the logarithm of the figures.

# 9

# Time Series

**9.1. Introduction.** In the preceding chapter, mention was made of time series, and it was shown how the method of least squares could be used to fit a straight line, or other mathematical formula, to a series of data which changed with time. *Business*, *economics*, *population studies*, and *weather* are typical of the areas where time series play a vital role for both *recording* and *forecasting*.

The full analysis of a time series is more complex than the simple curve fitting discussed in the previous chapter, because the data are complicated by cyclical and seasonal variations.

**9.2. Classification of Time Series Movements.** A typical time series is shown in Figure 9.1. It will be seen that while the graph shows an overall *trend* of increasing value with time, there is imposed upon the general trend other variations which, it will be noted, are related to the season of the year.

The *movements* of a time series may be classified as follows:

1. *Secular.* This is the *long term* growth or decline. The determination of this *trend* can be made by the method of least squares (Chapter 8), or by other means discussed in this chapter.

2. *Cyclical.* These are the oscillations, with duration greater than a year, imposed on the data due to swings in the *business cycle or other similar causes.*

3. *Seasonal.* These are the oscillations, which depend on the *season of the year.* Thus, employment is usually higher at harvest time in the country and before Christmas in the cities. Rainfall will be higher at some times of the year than at others. (In some statistics, such as temperature and height of tides, there are even shorter, *daily oscillations.*)

4. *Random.* Empirical time series will be subject also to

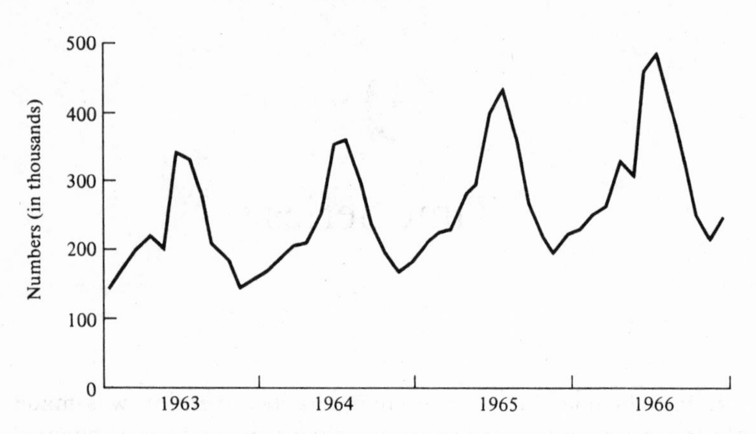


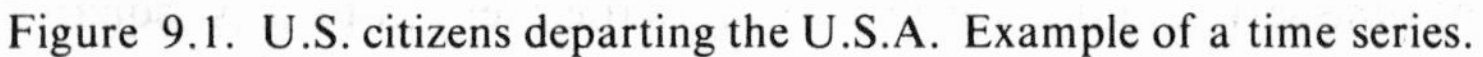
Figure 9.1. U.S. citizens departing the U.S.A. Example of a time series.

*chance*, and *random* or *irregular variations*. These movements are called *random* or sometimes *residual*.

In the study of time series, the secular or long term trend is usually of greatest interest, but this is not always so. A study of past seasonal movements is needed, for example, to interpret monthly production or other figures as they develop.

EXAMPLE 9.1. Give examples of causes which would lead to the four types of movement listed above.

*Solution.*

| | |
|---|---|
| *Secular* | Population growth |
| | Inflation of values |
| | Higher standard of living |
| *Cyclical* | Prosperity and slumps |
| | Changes in fashion |
| | Sun spot cycle (for weather statistics) |
| *Seasonal* | Changes in temperature |
| | Changes in weather |
| | Incidence of public and school holidays |
| *Random* | Fire or theft losses |
| | An unseasonal warm Sunday (for automobiles on the road) |
| | Sudden death of a President |
| | Declaration of War |

**9.3. Moving Averages.** In order to reduce the effect of random movements on time series data, we may substitute the *average value over a number of recordings*, for each individual value. Thus, if the series is $x_1, x_2, x_3 \ldots$, we could calculate

$$\frac{x_1 + x_2 + x_3}{3}, \frac{x_2 + x_3 + x_4}{3}, \text{etc.}$$

It will be noted that the first term corresponds in time to the individual value $x_2$, and hence, a moving average shortens the time series at each end. While this is normally unimportant at the beginning of the series, it does make the last recording less recent.

Moving averages using 3, 5, or some other odd number of terms have the advantage that the average figure corresponds to one of the original terms of the series, but in certain cases an even number of terms is more suitable. For example, when it is desired to remove seasonal movements, four quarters or twelve months moving averages should be used. (See next Section.)

EXAMPLE 9.2. The following data give the number of calls made by a salesman over a series of 30 working days.

| Day | Calls | Day | Calls | Day | Calls |
|---|---|---|---|---|---|
| 1 | 11 | 11 | 11 | 21 | 9 |
| 2 | 11 | 12 | 8 | 22 | 12 |
| 3 | 10 | 13 | 10 | 23 | 7 |
| 4 | 9 | 14 | 7 | 24 | 12 |
| 5 | 7 | 15 | 10 | 25 | 7 |
| 6 | 10 | 16 | 7 | 26 | 10 |
| 7 | 12 | 17 | 6 | 27 | 11 |
| 8 | 6 | 18 | 6 | 28 | 11 |
| 9 | 10 | 19 | 10 | 29 | 10 |
| 10 | 11 | 20 | 10 | 30 | 9 |

Plot the graph of these data and of the five day moving average.
*Solution.*

| Day | Calls | 5 Day Moving Average | Day | Calls | 5 Day Moving Average |
|---|---|---|---|---|---|
| 1 | 11 | | 5 | 7 | 9.6 |
| 2 | 11 | | 6 | 10 | 8.8 |
| 3 | 10 | 9.6 | 7 | 12 | 9.0 |
| 4 | 9 | 9.4 | 8 | 6 | 9.8 |

| Day | Calls | 5 Day Moving Average | Day | Calls | 5 Day Moving Average |
|---|---|---|---|---|---|
| 9 | 10 | 10.0 | 20 | 10 | 9.4 |
| 10 | 11 | 9.2 | 21 | 9 | 9.6 |
| 11 | 11 | 10.0 | 22 | 12 | 10.0 |
| 12 | 8 | 9.4 | 23 | 7 | 9.4 |
| 13 | 10 | 9.2 | 24 | 12 | 9.6 |
| 14 | 7 | 8.4 | 25 | 7 | 9.4 |
| 15 | 10 | 8.0 | 26 | 10 | 10.2 |
| 16 | 7 | 7.2 | 27 | 11 | 9.8 |
| 17 | 6 | 7.8 | 28 | 11 | 10.2 |
| 18 | 6 | 7.8 | 29 | 10 | |
| 19 | 10 | 8.2 | 30 | 9 | |

**9.4. Date and Period Data.** Data for a time series may be at individual dates, for example, the Dow Jones stock market average on the first of each month; or for a week, month, quarter, or year, for example, the sales of the XYZ company each month. Moving averages will follow the pattern of the original data. Thus, the three month moving average of the price of a stock, using data for January 1, February 1, and March 1, will correspond to February 1. The four month moving average of the price using January 1, February 1, March 1, and April 1 will correspond to February 15th. The three month moving average based on sales for the month of January, February, and March will correspond to the month of February.

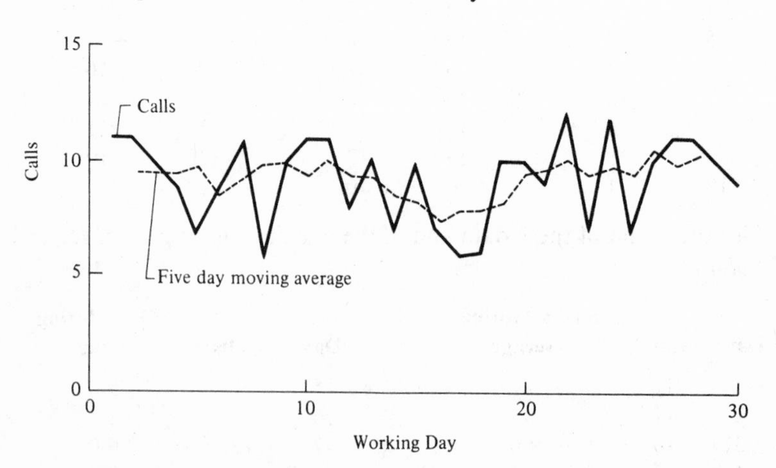


Figure 9.2. Salesman's daily calls. Five-day moving average.

If production and similar data are recorded in quarters of a year, and moving average figures corresponding to quarter years are required, the following formula should be used.

Moving average for quarter

$$= \frac{1}{8}(Q_{-2} + 2Q_{-1} + 2Q_0 + 2Q_1 + Q_2)$$

where $Q$ is the actual quarter's figures and the suffix represents the quarter to be used. If the moving average for the first quarter of any year is required, $Q_0$ = the first quarter's actual figures, $Q_1$ = the second quarter's actual figures, $Q_{-1}$ = the last quarter's actual figures for the preceding year, etc.

The corresponding formula for monthly moving averages is

$$\frac{1}{24}\left(M_{-6} + 2\sum_{x=-5}^{5} M_x + M_6\right)$$

It should be noted that these formulas produce the same results as taking the mean of two neighboring moving averages, e.g.,

$$\frac{1}{8}(Q_{-2} + 2Q_{-1} + 2Q_0 + 2Q_1 + Q_2)$$

$$= \frac{1}{2}\left[\frac{1}{4}(Q_{-2} + Q_{-1} + Q_0 + Q_1) + \frac{1}{4}(Q_{-1} + Q_0 + Q_1 + Q_2)\right]$$

EXAMPLE 9.3. Develop six day moving averages for the data in Example 9.2 to correspond with actual days by the method of the previous section.

*Solution.* The first figure will correspond to *day 4* and will be

$$\frac{1}{12}[\text{Day } 1 + 2\,(\text{Sum of Days 2 to 6}) + \text{Day } 7]$$

| Day | 6 Day Average | Day | 6 Day Average | Day | 6 Day Average |
|---|---|---|---|---|---|
| 4 | 9.8 | 12 | 9.5 | 20 | 8.9 |
| 5 | 9.4 | 13 | 9.2 | 21 | 9.5 |
| 6 | 9.0 | 14 | 8.4 | 22 | 9.8 |
| 7 | 9.2 | 15 | 7.8 | 23 | 9.5 |
| 8 | 9.7 | 16 | 7.7 | 24 | 9.7 |
| 9 | 9.8 | 17 | 7.9 | 25 | 9.8 |
| 10 | 9.5 | 18 | 8.1 | 26 | 9.9 |
| 11 | 9.4 | 19 | 8.4 | 27 | 9.9 |

**9.5. Methods of Determining Trends.** There are four principal methods of determining the trend in a time series. These are:

1. *Graphical.* The simplest method of determining a trend is to plot the data on a graph and then to *draw in free hand a smooth curve* which *cuts through the seasonal and other short term variations* and follows the *general trend.* If the curve is approximately a straight line, a ruler can be used to draw it. Flexible rules, which are available from art supply shops, are useful when the data do not approximate to a straight line.

*Advantages.*
(1) Simple
(2) Not bound to any mathematical formula
(3) Excellent results *if done with skill*

*Disadvantages.*
(1) Requires considerable practice and skill for good results
(2) Depends on personal judgment

2. *Semi-Average.* This method is *appropriate only when the trend is approximately a straight line.* The data are divided into two equal (or approximately equal) time ranges. The average value of each range is calculated and plotted at the center point, in time, of each half. These two points are then joined by a straight line which is extended in each direction to span the whole time range.

*Advantages.*
(1) Simple
(2) Free from judgment
(3) Suitable for forecasting

*Disadvantages.*
(1) Only suitable for a straight line trend
(2) Not as satisfactory as the least squares method mentioned below
(3) Must be recalculated as new data becomes available

3. *Moving Average.* This method has been described previously in Section 9.3. This method smooths to a limited extent only and provides no definite trend curve for future extrapolation.

*Advantages.*
(1) Simple
(2) Free from judgment (except in the selection of number of periods averaged)

(3) Can be extended when additional data becomes available without any recalculation

*Disadvantages.* (1) Not a completely smooth curve
(2) Does not provide a smooth curve over the whole range (ends short of the latest date for which data are available)

4. *Least Squares.* This method is somewhat similar, but more scientific than the semi-average method and can be used for both straight lines and more complicated trend formulas. The method was described in Chapter 8.

*Advantages.* (1) Free from judgment
(2) Provides the "best fitting curve"
(3) Suitable for forecasting

*Disadvantages.* (1) Results depend on the assumption of a definite mathematical curve.
(2) The method is rather laborious unless data processing equipment is available
(3) The method gives too much weight to abnormal values

EXAMPLE 9.4. Apply the graphic method to the following data for the sales of a certain periodical over a period of 24 quarters.

| Quarter | Sales in Thousands | Quarter | Sales in Thousands | Quarter | Sales in Thousands |
|---|---|---|---|---|---|
| 1 | 60 | 9 | 90 | 17 | 130 |
| 2 | 70 | 10 | 100 | 18 | 140 |
| 3 | 50 | 11 | 120 | 19 | 130 |
| 4 | 80 | 12 | 110 | 20 | 140 |
| 5 | 60 | 13 | 120 | 21 | 150 |
| 6 | 80 | 14 | 130 | 22 | 150 |
| 7 | 80 | 15 | 140 | 23 | 140 |
| 8 | 90 | 16 | 140 | 24 | 150 |

*Solution.* The results of the graphic determination of the trend are shown in Figure 9.3. Note that the judgment determination has been made to show the trend as a curve and not as a straight line.

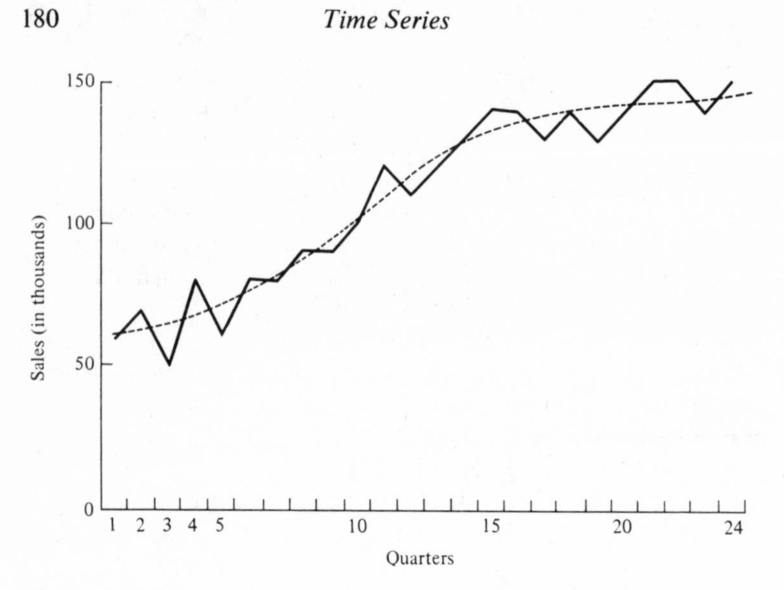


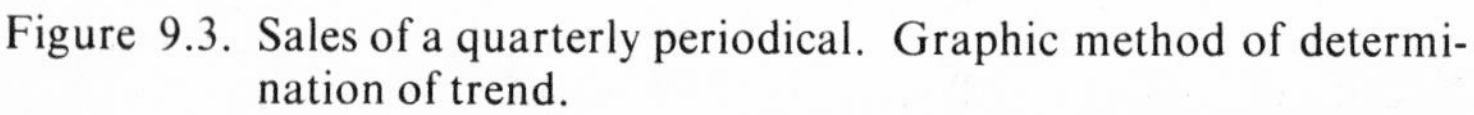
Figure 9.3. Sales of a quarterly periodical. Graphic method of determination of trend.

EXAMPLE 9.5. Apply the semi-average method to determine the trend of the data in Example 9.4.

*Solution.*

| Quarter | Sales in Thousands | Quarter | Sales in Thousands |
|---|---|---|---|
| 1 | 60 | 13 | 120 |
| 2 | 70 | 14 | 130 |
| 3 | 50 | 15 | 140 |
| 4 | 80 | 16 | 140 |
| 5 | 60 | 17 | 130 |
| 6 | 80 | 18 | 140 |
| 7 | 80 | 19 | 130 |
| 8 | 90 | 20 | 140 |
| 9 | 90 | 21 | 150 |
| 10 | 100 | 22 | 150 |
| 11 | 120 | 23 | 140 |
| 12 | 110 | 24 | 150 |
| | 990 | | 1,660 |
| Average (÷12) = | 83 | | = 138 |

These averages correspond to $6\frac{1}{2}$ and $18\frac{1}{2}$ respectively, the midpoints of the two ranges. The result is shown in Figure 9.4.

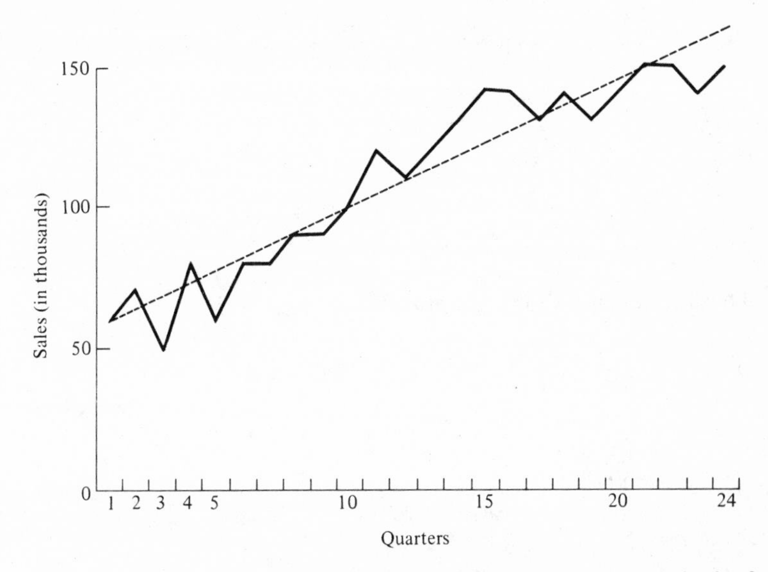


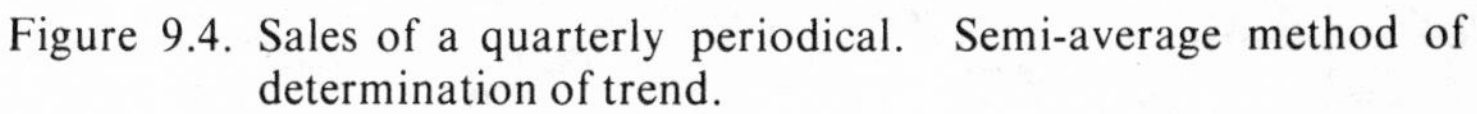

Figure 9.4. Sales of a quarterly periodical. Semi-average method of determination of trend.

EXAMPLE 9.6. Calculate the 4 quarter moving average to determine the trend of the data in Example 9.4.
*Solution.*

| Quarter | Sales in Thousands | 4 Period Total | 4 Period Average |
|---|---|---|---|
| 1 | 60 | | |
| 2 | 70 | 260 | 65 |
| 3 | 50 | 260 | 65 |
| 4 | 80 | 270 | 68 |
| 5 | 60 | 300 | 75 |
| 6 | 80 | 310 | 78 |
| 7 | 80 | 340 | 85 |
| 8 | 90 | 360 | 90 |
| 9 | 90 | 400 | 100 |
| 10 | 100 | 420 | 105 |
| 11 | 120 | 450 | 113 |
| 12 | 110 | 480 | 120 |
| 13 | 120 | 500 | 125 |
| 14 | 130 | 530 | 133 |
| 15 | 140 | 540 | 135 |
| 16 | 140 | 550 | 138 |
| 17 | 130 | 540 | 135 |

| Quarter | Sales in Thousands | 4 Period Total | 4 Period Average |
|---|---|---|---|
| 18 | 140 | | |
| | | 540 | 135 |
| 19 | 130 | | |
| | | 560 | 140 |
| 20 | 140 | | |
| | | 570 | 143 |
| 21 | 150 | | |
| | | 580 | 145 |
| 22 | 150 | | |
| | | 590 | 148 |
| 23 | 140 | | |
| 24 | 150 | | |

These results are shown graphically in Figure 9.5.

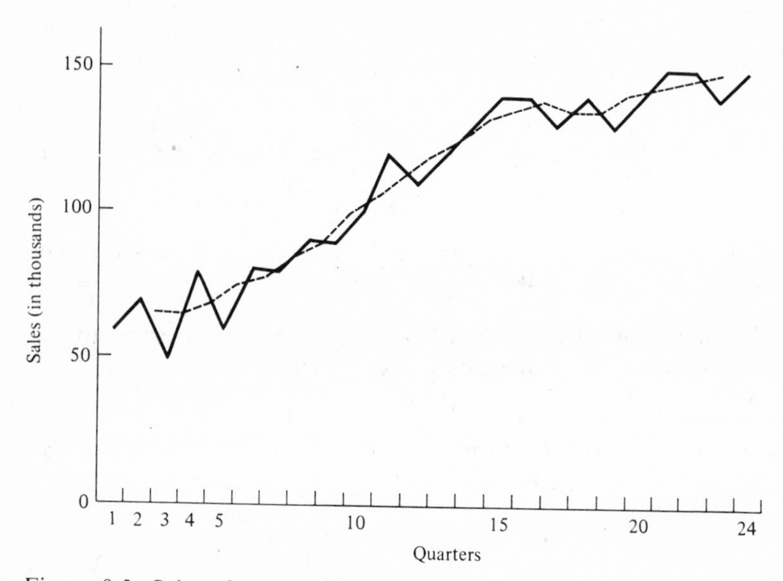


Figure 9.5. Sales of a quarterly periodical. Moving average method of determination of trend.

**9.6. Annual Rate of Growth.** In many business and similar time series statistics, considerable importance is attached to the rate of growth. For a year $x$, the rate of growth over the previous year is

$$\frac{f(x) - f(x - 1)}{f(x - 1)} \tag{9.1}$$

where $f(x)$ is the value of the function. Such rates of growth are usually expressed as a percentage. Thus, if sales in 1966 are

\$90,000 compared with \$80,000 in 1965, the rate of growth of sales in 1966 is

$$\frac{90{,}000 - 80{,}000}{80{,}000} = .125 \text{ or } 12\tfrac{1}{2}\%$$

It should be noted that if sales (say) followed an exact trend line which was a straight line, the rate of growth over a long period of time would fall. An *exponential* growth line is needed to maintain a steady rate of growth.

EXAMPLE 9.7. The trend line developed in the solution to Example 9.5 was a straight line through the points ($6\tfrac{1}{2}$, 83)($18\tfrac{1}{2}$, and 138). Obtain a formula for this trend and calculate the *quarterly* rate of growth for the first quarter and the last (24th) quarter, of the trend line figures.

*Solution.* The formula for a straight line is

$$y = ax + b$$

Substituting the values given,

$$83 = a \times 6\tfrac{1}{2} + b$$
$$138 = a \times 18\tfrac{1}{2} + b$$

Subtracting,

$$55 = 12a$$
$$a = 4.58$$

and

$$b = 83 - 4.58 \times 6\tfrac{1}{2}$$
$$b = 53.23$$

The trend line is

$$y = 4.58x + 53.23$$

When

$$x = 0, \qquad y = 53.23$$
$$x = 1, \qquad y = 57.81$$

The quarterly rate of growth for the second quarter is

$$\frac{f(1) - f(0)}{f(0)}$$

which equals

$$\frac{57.81 - 53.23}{53.23} = \frac{4.58}{53.23} = .086 \text{ or } 8.6\%$$

The quarterly rate of growth for the last quarter is

$$\frac{f(24) - f(23)}{f(23)}$$

which equals

$$\frac{163.15 - 158.57}{158.57} = \frac{4.58}{158.57} = 0.29 \text{ or } 2.9\%$$

**9.7. Seasonal Variation.** Given a volume of monthly data, it may be found that some months of the year are regularly higher or lower than the general average due to seasonal variations. A *seasonal index* is a factor which indicates the average departure from the trend line for any month of the year. Each month will have an index which will be greater than 1 for months with high values of the data and will be less than 1 for months with low values. The average seasonal index for all months should be 1, and hence the total of the seasonal indexes for all the twelve months of the year should be 12. Seasonal indexes are sometimes expressed as percentages. While the above remarks, and what follows, assume monthly data, quarterly and weekly indexes are sometimes used.

If there were no trend, such indexes would be calculated readily by adding up the data for all Januaries, all Februaries, etc., and dividing by the number of years to obtain typical monthly values. Dividing the average value for, say, January, by the average value for all months combined would give the January seasonal index, and similarly for other months.

However, if there is an overall upward trend in data recorded over a series of calendar years, the average value for December will be greater than the average value for January *for this reason alone*. Hence, when trends are involved, an adjustment for trend must be introduced to allow for this in the calculation of seasonal indexes. Four, rather similar, methods are used for calculating seasonal indexes where trends are involved. These are described below.

**9.8. Calculation of Seasonal Index.**

1. *Average Method.* The procedure described above is carried through and then the February index is divided by 1/12 the annual trend, the March index by 2/12 the annual trend, and so on. The adjusted trend figure will not add up to 12, and each

will be increased or decreased by the ratio of 12 divided by their total to correct this.

*Modifications.* (1) It is often preferable to express all monthly figures as a percentage of the average monthly figure for the year, before starting the calculations.

(2) Instead of calculating the mean monthly figure for each month, the median is sometimes preferred, omitting unusual extreme values which are probably due to random variation.

2. *Trend Method.* Starting with the trend value, which has previously been calculated (or read off a graph) for each month, the data for each month is expressed as a percentage of the trend value. The averages of the percentages for the respective months then give the required index. As in method 1, if they do not add up to 12 they should be adjusted appropriately.

3. *Moving Average Method.* This method is similar to the trend method, except that the *moving average* is substituted for the trend value.

4. *Link Related Method.* Each month's data can be expressed as a percentage of the previous month. These percentages are links. These links are first calculated for each month's data and then the average February–January link, the average March–February link, etc., are calculated. Suppose the links are:

| | |
|---|---|
| February–January | $l_1$ |
| March–February | $l_2$ |
| .... | |
| January–December | $l_{12}$ |

$l_1 \times l_2 \times l_3 \times \cdots \times l_{12}$ is then calculated. If the data have a trend, this multiplication will not produce unity but the annual trend, say $1 + k$. Each value of $l$ must be multiplied by $1 - k/12$ to adjust for the trend. The adjusted values of $l$ can then be multiplied in succession to produce a series of indexes of seasonal variation based on January as 1, thus

| | |
|---|---|
| January | 1 |
| February | $l_1$ |
| March | $l_1 \times l_2$ |
| .... | |
| December | $l_1 \times l_2 \times \cdots \times l_{11}$ |

If necessary the indexes must be increased appropriately to sum to 12.

All four methods normally produce closely similar results, but the actual results will differ because of cyclical movements in the data being treated somewhat differently. No particular method is to be preferred over the others.

EXAMPLE 9.8. The data for U.S. citizens departing the U.S.A. for foreign travel is shown graphically in Figure 9.1. The actual figures are as follows. All numbers are in thousands.

| Month | 1963 | 1964 | 1965 | 1966 | Four Years |
|---|---|---|---|---|---|
| January | 142 | 172 | 207 | 232 | 753 |
| February | 176 | 193 | 225 | 248 | 842 |
| March | 201 | 206 | 234 | 262 | 903 |
| April | 212 | 214 | 278 | 330 | 1034 |
| May | 200 | 253 | 296 | 308 | 1057 |
| June | 341 | 356 | 398 | 459 | 1554 |
| July | 333 | 359 | 433 | 486 | 1611 |
| August | 283 | 302 | 365 | 396 | 1346 |
| September | 207 | 238 | 265 | 322 | 1032 |
| October | 186 | 195 | 224 | 250 | 855 |
| November | 146 | 167 | 195 | 217 | 725 |
| December | 161 | 186 | 221 | 248 | 816 |
| | 2588 | 2841 | 3341 | 3758 | 12528 |

Calculate the trend line by the semi-average method, and calculate the monthly seasonal indexes for the data by the *average method.*

*Solution.* Average values for the first and second halves of the experience are

$$\frac{2588 + 2841}{24} \quad \text{and} \quad \frac{3341 + 3758}{24}$$

or,

$$226.2 \quad \text{and} \quad 295.8$$

These correspond to January 1, 1964 and January 1, 1966 respectively, a two-year period.

The annual trend in monthly figures expressed in thousands of travelers is

$$\frac{295.8 - 226.2}{2} = \frac{69.6}{2} = 34.8$$

The average monthly value for the whole 4 years is

$$\frac{2588 + 2841 + 3341 + 3758}{48} = 261$$

The annual trend, as a ratio, is 34.8/261, or 13.3%, and the monthly trend is 13.3%/12 = 1.1%.

The calculation of the seasonal indexes proceeds as follows.

| Month (1) | Total 4 Years (2) | Typical Month (3) | Unadjusted Seasonal Index (4) | Trend Adj. (5) | (4) ÷ (5) (6) | Seasonal Index Adj. for Trend (7) |
|---|---|---|---|---|---|---|
| Jan. | 753 | 188 | .72 | 1.000 | .72 | .76 |
| Feb. | 842 | 210 | .80 | 1.011 | .79 | .84 |
| Mar. | 903 | 226 | .87 | 1.022 | .85 | .90 |
| Apr. | 1034 | 259 | .99 | 1.033 | .96 | 1.02 |
| May | 1057 | 264 | 1.01 | 1.044 | .97 | 1.03 |
| Jun. | 1554 | 388 | 1.49 | 1.055 | 1.41 | 1.49 |
| July | 1611 | 403 | 1.54 | 1.066 | 1.44 | 1.53 |
| Aug. | 1346 | 337 | 1.29 | 1.077 | 1.20 | 1.27 |
| Sept. | 1032 | 258 | .99 | 1.088 | .91 | .96 |
| Oct. | 855 | 214 | .82 | 1.099 | .75 | .80 |
| Nov. | 725 | 181 | .69 | 1.110 | .62 | .66 |
| Dec. | 816 | 204 | .78 | 1.121 | .70 | .74 |
| | 12,528 | 3,132 | | | 11.32 | 12.00 |

*Notes.* Column (3) is column (2) ÷ 4.
Column (4) is column (3) divided by the monthly average 3132 ÷ 12 = 261. This column could have been calculated by dividing column (2) by the column total ÷ 12.
Column (5) is $[1 + .011 \times (t - 1)]$ where $t$ is the month.
Column (7) is column (6) increased in the ratio 12 ÷ 11.32.

EXAMPLE 9.9. Using the data in Example 9.8, and the trend developed in the solution of that example, calculate the monthly seasonal indexes for the data by the *trend method.*

*Solution.* The two points for the trend value may be plotted, and values for individual months read off. Alternatively, they may be calculated by the formula for a straight line. If the midpoint of January 1963 is represented by $x = 0$, the midpoint for February 1963 represented by $x = 1$, etc., the two points calculated are $x = 11\frac{1}{2}$ and $x = 35\frac{1}{2}$.

The formula for a straight line is $y = a + bx$. Solving for $a$ and $b$ gives

$$y = 192.85 + 2.9x$$

The trend figures can now be calculated and the actual and trend figures tabulated.

| | 1963 | | 1964 | | 1965 | | 1966 | | Average Monthly Percent-age | Seasonal* Index |
|---|---|---|---|---|---|---|---|---|---|---|
| **Month** | **Trend** | **A/T†** | **Trend** | **A/T** | **Trend** | **A/T** | **Trend** | **A/T** | | |
| Jan. | 193 | 74% | 228 | 75% | 262 | 79% | 297 | 78% | 77% | 77% |
| Feb. | 196 | 90 | 231 | 84 | 265 | 85 | 300 | 83 | 86 | 86 |
| Mar. | 199 | 101 | 233 | 88 | 268 | 87 | 303 | 86 | 90 | 90 |
| Apr. | 202 | 105 | 236 | 91 | 271 | 103 | 306 | 108 | 102 | 102 |
| May | 204 | 98 | 239 | 106 | 274 | 108 | 309 | 100 | 103 | 102 |
| Jun. | 207 | 165 | 242 | 147 | 277 | 144 | 312 | 147 | 151 | 150 |
| July | 210 | 159 | 245 | 147 | 280 | 155 | 315 | 154 | 154 | 153 |
| Aug. | 213 | 133 | 248 | 122 | 283 | 129 | 318 | 125 | 127 | 126 |
| Sept. | 216 | 96 | 251 | 95 | 286 | 93 | 320 | 101 | 96 | 96 |
| Oct. | 219 | 85 | 254 | 77 | 289 | 78 | 323 | 77 | 79 | 79 |
| Nov. | 222 | 66 | 257 | 65 | 291 | 67 | 326 | 67 | 66 | 66 |
| Dec. | 225 | 72 | 260 | 72 | 294 | 75 | 329 | 75 | 73 | 73 |
| | | | | | | | | | 1204 | 1200 |

†A/T = Actual divided by Trend.
*Average Monthly Percentage × 1200/1204.

It will be noted that the results are very close to those obtained in Example 9.8.

EXAMPLE 9.10. Use the data in Example 9.8, and the *moving average method* to determine the monthly seasonal index.

*Solution.* The moving average must be based on 12 month figures, because of the marked season variation. This will produce trend figures corresponding to the *first day of each month*. The mean of the first day of January and February must then be calculated to give the moving average trend for *January* and similarly for other months.

| Month | Increase 1964 over 1963 | Twelve Month Year Ending | Total |
|---|---|---|---|
| | | 12/31/63 | 2588 |
| Jan. | 30 | 1/31/64 | 2618 |
| Feb. | 17 | 2/28/64 | 2635 |
| Mar. | 5 | 3/31/64 | 2640 |
| Apr. | 2 | 4/30/64 | 2642 |
| May | 53 | 5/31/64 | 2695 |
| Jun. | 15 | 6/30/64 | 2710 |
| July | 26 | 7/31/64 | 2736 |
| Aug. | 19 | 8/31/64 | 2755 |
| Sept. | 31 | 9/30/64 | 2786 |
| Oct. | 9 | 10/31/64 | 2795 |
| Nov. | 21 | 11/30/64 | 2816 |
| Dec. | 25 | 12/31/64 | 2841 |

| Month | Monthly Trend Value | Actual | A/T |
|---|---|---|---|
| July 63 | 217 | 333 | 153% |
| Aug. 63 | 219 | 283 | 129 |
| Sept. 63 | 220 | 207 | 94 |
| Oct. 63 | 220 | 186 | 85 |
| Nov. 63 | 222 | 146 | 66 |
| Dec. 63 | 225 | 161 | 72 |
| Jan. 64 | 227 | 172 | 76 |
| Feb. 64 | 229 | 193 | 84 |
| Mar. 64 | 231 | 206 | 89 |
| Apr. 64 | 233 | 214 | 92 |
| May 64 | 234 | 253 | 108 |
| Jun. 64 | 236 | 356 | 151 |

Similarly for the rest of the table. These results are then brought together in a summary and the final seasonal index calculated.

| | Actual Divided by Trend | | | | | |
|---|---|---|---|---|---|---|
| Month | 1963 | 1964 | 1965 | 1966 | Average Monthly Percentage | Seasonal Index |
| Jan. | –% | 76% | 80% | 78% | 78% | 78% |
| Feb. | – | 84 | 85 | 82 | 84 | 84 |
| Mar. | – | 89 | 87 | 86 | 87 | 87 |
| Apr. | – | 92 | 102 | 107 | 100 | 100 |
| May | – | 108 | 108 | 99 | 105 | 105 |
| Jun. | – | 151 | 144 | 147 | 147 | 148 |
| July | 153 | 151 | 155 | – | 153 | 154 |
| Aug. | 129 | 125 | 130 | – | 128 | 129 |
| Sept. | 94 | 98 | 93 | – | 95 | 95 |
| Oct. | 85 | 79 | 78 | – | 81 | 81 |
| Nov. | 66 | 66 | 67 | – | 66 | 66 |
| Dec. | 72 | 73 | 75 | – | 73 | 73 |
| | | | | | 1197 | 1200 |

EXAMPLE 9.11. Use the data in Example 9.8, and the *Link Related Method* to determine the monthly seasonal index.

*Solution.* The links are first calculated and then averaged.

| | **1963** | **1964** | **1965** | **1966** | **Average** |
|---|---|---|---|---|---|
| Feb./Jan. | 124% | 112% | 109% | 107% | 113% |
| Mar./Feb. | 114 | 107 | 104 | 106 | 108 |
| Apr./Mar. | 105 | 104 | 119 | 126 | 114 |
| May/Apr. | 94 | 118 | 106 | 93 | 103 |
| Jun./May | 171 | 141 | 134 | 149 | 149 |
| Jul./Jun. | 98 | 101 | 109 | 106 | 103 |
| Aug./Jul. | 85 | 84 | 84 | 81 | 83 |
| Sept./Aug. | 73 | 79 | 73 | 81 | 77 |
| Oct./Sept. | 90 | 82 | 85 | 78 | 84 |
| Nov./Oct. | 78 | 86 | 87 | 87 | 84 |
| Dec./Nov. | 110 | 112 | 113 | 114 | 112 |
| Jan./Dec. | 107 | 112 | 105 | – | 108 |

$$l_1 \times l_2 \times l_3 \times \cdots \times l_{12} = 1.20 = 1 + k$$

$$k = 0.20$$

$$\text{and} \quad \frac{k}{12} = 0.017$$

$$1 - \frac{k}{12} = 0.983$$

Therefore, the links must be multiplied by 0.983 to remove seasonal trend and the calculation proceeds as follows.

| **Month** | **Link** | **Link Adjustment** | **Adj. Link** | **Accumulated Links** | **Monthly Index** |
|---|---|---|---|---|---|
| Jan. | | | | 100% | 78% |
| Feb. | 113% | .983 | 111% | 111 | 86 |
| March | 108 | .983 | 106 | 118 | 92 |
| April | 114 | .983 | 112 | 132 | 102 |
| May | 103 | .983 | 101 | 133 | 103 |
| June | 149 | .983 | 146 | 194 | 151 |
| July | 103 | .983 | 101 | 196 | 152 |
| Aug. | 83 | .983 | 82 | 161 | 125 |
| Sept. | 77 | .983 | 76 | 122 | 95 |
| Oct. | 84 | .983 | 83 | 102 | 79 |
| Nov. | 84 | .983 | 83 | 84 | 65 |
| Dec. | 112 | .983 | 110 | 93 | 72 |
| | | | | 1546 | 1200 |

**9.9. Seasonally Adjusted Data.** Many important national indexes such as production, unemployment, etc., are subject to appreciable seasonal variation. In order to keep a careful watch on cyclical movements, it is necessary to calculate seasonally adjusted figures. Such adjusted figures will include trend, cyclical, and random movements.

**9.10. Estimating of Cyclical and Random Movements.** Data which has been adjusted for both trend and seasonal movements will reveal the cyclical movements as longer term variations and random movements as short variations often limited to a single month.

**9.11. Comparison of Monthly Data.** In comparing monthly production and similar data, it is important to remember that the number of working days in a month may vary appreciably owing to the exact length of the month and the incidence of Sundays and holidays. It is often useful to calculate in such circumstances the *production per working day*.

## Problems

**Problem 9.1.** Classify the movements you would expect in a time series from the following causes.

(1) An earthquake
(2) A period of deflation
(3) The football season
(4) Tornados (as they affect statistics for an individual neighborhood)
(5) Tornados (as they affect statistics for the whole Middle West of the U.S.)
(6) Expansion of college enrollment

**Problem 9.2.** The normal seasonal cycle is one year. Give two examples of shorter cycles which could occur in statistics and state the period of each.

**Problem 9.3.** Is it true to say that the longer the period used in calculating a moving average, the greater the smoothing? If not, what period is the best for smoothing?

**Problem 9.4.** The following table gives the value of exports of U.S. merchandise by quarter over the period 1963 to 1966. Smooth these data by calculating the four-quarter moving averages.

| Year | QUARTER 1 | 2 | 3 | 4 |
|---|---|---|---|---|
| 1963 | 115 | 134 | 122 | 142 |
| 1964 | 142 | 146 | 138 | 162 |
| 1965 | 128 | 163 | 148 | 172 |
| 1966 | 162 | 169 | 161 | 180 |

(Values are based on an index of 100 for the period 1957–59.)

Compare the smoothness of the original and averaged data by finding the maximum change from one quarter to the next in both cases.

**Problem 9.5.** The moving averages calculated in the previous problem corresponded to periods halfway between the periods of the original data. Calculate four-quarter moving averages adjusted to agree with the original period.

**Problem 9.6.** The following were the number of Coast Guard personnel on active duty at June 30th in the years indicated.

| Year | 1940 | 1945 | 1950 | 1955 | 1960 | 1965 |
|---|---|---|---|---|---|---|
| Number | 13,756 | 171,192 | 23,190 | 28,607 | 30,616 | 31,776 |

To what extent are the variations indicated secular, cyclical, seasonal, and random?

**Problem 9.7.** For the following data, which are recorded at a certain date and which for a period of time?

(1) Sales of a departmental store.
(2) The Dow Jones average stock price.
(3) The population of the U.S.A.
(4) Unemployment figures.
(5) Company salary figures.
(6) School enrollment.

**Problem 9.8.** The following data give the enrollment in thousands in institutions of higher education.

| Year | 1930 | 1940 | 1950 | 1960 | 1965 |
|---|---|---|---|---|---|
| Number | 1,101 | 1,494 | 2,659 | 3,216 | 5,500 |

Plot these data and draw a line fitting them. Explain why all the increase cannot be attributed to secular trend.

**Problem 9.9.** Fit a least squares curve to the data in Problem 9.8, and compare it to the graphic fitted curve in the answer to Problem 9.8.

**Problem 9.10.** Calculate by the semi-average method, the straight line trend of the data in Problem 9.8. Compare this result with the answer to Problem 9.9.

**Problem 9.11.** Which of the four methods for determining trend described in this chapter can be used when the data are not regularly spaced on the time scale?

**Problem 9.12.** If a certain function has a value of 1,000 at time 0 and an annual rate of growth of 10%, what is

(1) a formula for the function at time $t$
(2) the value of the function at time $t = \frac{1}{2}$
and at time $t = 10$ years?

**Problem 9.13.** The rate of growth for sales, and similar statistics recorded on a monthly basis, is usually in the form of the percentage increase over the *corresponding month for the preceding year.* This is an *annual* rate of growth, since it compares figures for one year with figures for a year previous. It has the advantage of avoiding the difficulties due to seasonal fluctuations. If data were free from all seasonal fluctuations, monthly rates of growth could be calculated by comparing one month with the preceding month. If such a comparison showed a steady monthly rate of growth of 1% what would be the annual rate of growth?

**Problem 9.14.** If a certain statistic was not subject to appreciable variation from day to day, monthly figures would vary because of the different number of days in the month. What would be the *seasonal index* for such data for

(1) January
(2) February (for a year which is not a leap year)?

**Problem 9.15.** The following table shows, for the U.S.A., profits and dividends from all manufacturing corporations by calendar quarter, in billions of dollars.

| | QUARTER | | | |
|---|---|---|---|---|
| **Year** | **1** | **2** | **3** | **4** |
| 1963 | 4.0 | 5.2 | 4.8 | 5.5 |
| 1964 | 5.1 | 6.1 | 5.7 | 6.3 |
| 1965 | 6.2 | 7.2 | 6.6 | 7.5 |
| 1966 | 7.2 | 8.4 | 7.4 | 7.9 |

Use the Average Method to calculate seasonal indexes for these data.

**Problem 9.16.** Use the Trend Method to calculate seasonal indexes for the data in Problem 9.15.

**Problem 9.17.** Use the Moving Average Method to calculate seasonal indexes for the data in Problem 9.15.

**Problem 9.18.** Use the Link Method to calculate seasonal indexes for the data in Problem 9.15.

**Problem 9.19.** The data in Problem 9.15 has been analysed for trend and seasonal variation. The residual variation must be due to cyclical or random causes. In which quarter is this residual variation greatest?

**Problem 9.20.** The following figures give production and working days in a 6 month period.

| Month | Production | Working Days |
|---|---|---|
| 1 | 8540 | 22 |
| 2 | 7415 | 19 |
| 3 | 8265 | 21 |
| 4 | 8274 | 21 |
| 5 | 9210 | 23 |
| 6 | 8107 | 20 |

Calculate the production per working day for each month. Which month was most productive per working day?

## Solutions

**Problem 9.1.**

(1) Random
(2) Cyclical
(3) Seasonal
(4) Random
(5) Seasonal
(6) Secular

**Problem 9.2.**

(1) Travel on public transportation—weekly cycle.
(2) Work load in shops, on telephone exchanges, etc.,
—daily cycle.

**Problem 9.3.** If data are subject to only random variations, the longer the period used in calculating the moving average, the greater the smoothing. However, where seasonal variations are predominant, a period equal to, or a multiple of, the seasonal period is best for smoothing.

**Problem 9.4.** Four-quarter moving averages are

128, 135, 138, 142, 147, 144, 148, 150, 153, 161, 163, 166, 168

The greatest change between successive quarters in the original data is 163 − 128 = 35. The greatest change between moving averages is 161 − 153 = 8.

**Problem 9.5.** For this purpose we use the formula

$$\frac{1}{8}\,(Q_{-2} + 2Q_{-1} + 2Q_0 + 2Q_1 + Q_2)$$

or we can take the mean of adjacent figures calculated in the previous solution.

| | **QUARTER** | | | |
|---|---|---|---|---|
| **Year** | **1** | **2** | **3** | **4** |
| 1963 | | | 132 | 137 |
| 1964 | 140 | 145 | 145 | 146 |
| 1965 | 149 | 152 | 157 | 162 |
| 1966 | 164 | 167 | | |

**Problem 9.6.** Since the figures given are at June 30th in quinquennial years, no seasonal movements are involved. The high value in 1945, due to the World War, is random. The steady increase from 1940 to 1965 (ignoring the 1945 figures) is partly secular, and we trust, partly cyclical due to the Korean and Vietnam wars. There is no way to distinguish between the two from the data given.

**Problem 9.7.**

(1) Sales of a department store will be weekly, monthly, or yearly totals.

(2) Dow Jones average stock price will be the average closing price at the end of a particular day.

(3) The population of the U.S.A. will be as tabulated on a census date and estimated on other dates.

(4) Unemployment figures. These figures can be on a particular date or the average for a month or year. The latter is more usual.

(5) Company salary figures are normally monthly or yearly.

(6) School enrollment. This is usually at the beginning of a term, that is, at a fixed date.

**Problem 9.8.** College attendance increase is due to

(1) Population growth—secular.
(2) Birthrate changes some 20 years earlier—cyclical.
(3) Proportion of young people attending college—secular. (However, the present increase cannot continue indefinitely.)

**Problem 9.9.**

| Year | x | y | $x^2$ | xy | Fitted Values |
|---|---|---|---|---|---|
| 1930 | −2 | 1101 | 4 | −2202 | 670 |
| 1940 | −1 | 1494 | 1 | −1494 | 1788 |
| 1950 | 0 | 2659 | 0 | 0 | 2906 |
| 1960 | 1 | 3216 | 1 | 3216 | 4024 |
| 1965 | 1.5 | 5500 | 2.25 | 8250 | 4583 |
| | −0.5 | 13970 | 8.25 | 7770 | |

$$13970 = 5a_0 - 0.5a_1$$
$$7770 = -0.5a_0 + 8.25a_1$$

giving,

$$y = 2906 + 1118x$$

The fitted values are shown above.

**Problem 9.10.** Using 1950 as origin and 10 year intervals (as in the solution to Problem 9.9)

| Average first 3 dates x | y | | Average last 2 dates x | y |
|---|---|---|---|---|
| −2 | 1,101 | | 1 | 3,216 |
| −1 | 1,494 | | 1.5 | 5,500 |
| 0 | 2,659 | | | |
| −3 | 5,254 | *Total* | 2.5 | 8,716 |
| −1 | 1,751 | *Average* | 1.25 | 4,358 |

The straight line through these two points is

$$y = 2,910 + 1,159x$$

| | 1930 | 1940 | 1950 | 1960 | 1965 |
|---|---|---|---|---|---|
| **True Value** | 1101 | 1494 | 2659 | 3216 | 5500 |
| **Solution 9.9** | 670 | 1788 | 2906 | 4024 | 4583 |
| **Solution 9.10** | 592 | 1751 | 2910 | 4069 | 4648 |

**Problem 9.11.** The graphic, semi-average, and least squares can all be used for non-regular data. The moving-average requires evenly spaced data.

**Problem 9.12.** The formula for the function is

$$1{,}000 \times (1.1)^t$$

The values at the two points mentioned are

| t | value |
|---|---|
| $\frac{1}{2}$ | 1,049 |
| 10 | 2,594 |

**Problem 9.13.**

$$(1.01)^{12} - 1 = 12.7\%$$

but 1% × 12 or 12% is often used.

**Problem 9.14.** There are 365 days in an ordinary year, one-twelfth of this number is 30.417. January has 31 days, so that on the assumptions in the question, the seasonal index for January is

$$\frac{31}{30.417} = 1.019$$

The seasonal index for February is

$$\frac{28}{30.417} = .921$$

**Problem 9.15.** We must first calculate the trend line. The least squares method will be used.

| Year | Quarter | x | y | $x^2$ | xy |
|---|---|---|---|---|---|
| 1963 | 1 | −7 | −2.0 | 49 | +14.0 |
| | 2 | −6 | −0.8 | 36 | +4.8 |
| | 3 | −5 | −1.2 | 25 | +6.0 |
| | 4 | −4 | −0.5 | 16 | +2.0 |
| 1964 | 1 | −3 | −0.9 | 9 | +2.7 |
| | 2 | −2 | +0.1 | 4 | −0.2 |
| | 3 | −1 | −0.3 | 1 | +0.3 |
| | 4 | 0 | +0.3 | 0 | 0.0 |
| 1965 | 1 | 1 | +0.2 | 1 | +0.2 |
| | 2 | 2 | +1.2 | 4 | +2.4 |
| | 3 | 3 | +0.6 | 9 | +1.8 |
| | 4 | 4 | +1.5 | 16 | +6.0 |
| 1966 | 1 | 5 | +1.2 | 25 | +6.0 |
| | 2 | 6 | +2.4 | 36 | +14.4 |
| | 3 | 7 | +1.4 | 49 | +9.8 |
| | 4 | 8 | +1.9 | 64 | +15.2 |
| | | 8 | +5.1 | 344 | +85.4 |

$$5.1 = 16a_0 + 8a_1$$
$$85.4 = 8a_0 + 344a_1$$

Giving

$$a_0 = .197, \ a_1 = .244$$

The trend line is

$$(y - 6) = .197 + .244\,(x - 7)$$

where $x$ is in quarters from first quarter 1963 = 0.
The *annual* increase in quarterly amounts is .244 × 4 = .976
The average quarterly amount over the four years is

$$\frac{(6 \times 16) + 5.1}{16} = \frac{101.1}{16} = 6.32$$

The annual trend is

$$\frac{.976}{6.32} \quad \text{or} \quad 15.4\%$$

The quarterly trend is

$$\frac{15.4\%}{4} = 3.9\%$$

The calculation of the seasonal index is as follows:

| **Quarter** | **Total 4 Years** | **Typical Quarter** | **Unadjusted Seasonal Index** | **Trend Adj.** | **(4)÷(5)** | **Seasonal Index** |
|---|---|---|---|---|---|---|
| **(1)** | **(2)** | **(3)** | **(4)** | **(5)** | **(6)** | **(7)** |
| 1 | 22.5 | 5.6 | .89 | 1.00 | .89 | .94 |
| 2 | 26.9 | 6.7 | 1.06 | 1.04 | 1.02 | 1.09 |
| 3 | 24.5 | 6.1 | .97 | 1.08 | .90 | .95 |
| 4 | 27.2 | 6.8 | 1.08 | 1.12 | .96 | 1.02 |
| **Total** | 101.1 | 25.2 | 4.00 | | 3.77 | 4.00 |

**Problem 9.16.** From the trend formula in the solution to Problem 9.15, we can proceed as follows.

| Year | Quarter | Actual | Trend | Actual / Trend |
|---|---|---|---|---|
| 1963 | 1 | 4.0 | 4.49 | .89 |
| | 2 | 5.2 | 4.73 | 1.10 |
| | 3 | 4.8 | 4.98 | .96 |
| | 4 | 5.5 | 5.22 | 1.05 |
| 1964 | 1 | 5.1 | 5.47 | .93 |
| | 2 | 6.1 | 5.71 | 1.07 |
| | 3 | 5.7 | 5.95 | .96 |
| | 4 | 6.3 | 6.20 | 1.02 |
| 1965 | 1 | 6.2 | 6.44 | .96 |
| | 2 | 7.2 | 6.69 | 1.08 |
| | 3 | 6.6 | 6.93 | .95 |
| | 4 | 7.5 | 7.17 | 1.05 |
| 1966 | 1 | 7.2 | 7.42 | .97 |
| | 2 | 8.4 | 7.66 | 1.10 |
| | 3 | 7.4 | 7.91 | .94 |
| | 4 | 7.9 | 8.15 | .97 |

| Month | 1963 | 1964 | 1965 | 1966 | Total | Seasonal Index |
|---|---|---|---|---|---|---|
| 1 | .89 | .93 | .96 | .97 | 3.75 | .94 |
| 2 | 1.10 | 1.07 | 1.08 | 1.10 | 4.35 | 1.09 |
| 3 | .96 | .96 | .95 | .94 | 3.81 | .95 |
| 4 | 1.05 | 1.02 | 1.05 | .97 | 4.09 | 1.02 |
| | | | | | 16.00 | 4.00 |

**Problem 9.17.** The moving average will be calculated by the formula

$$\frac{1}{8}(Q_{-2} + 2Q_{-1} + 2Q_0 + 2Q_1 + Q_2)$$

to provide a slight variation from the solution in Example 9.10.

| Year | Quarter | Actual | Trend | Actual / Trend |
|---|---|---|---|---|
| 1963 | 1 | 4.0 | | |
| | 2 | 5.2 | | |
| | 3 | 4.8 | 5.01 | .96 |
| | 4 | 5.5 | 5.26 | 1.05 |

| Year | Quarter | Actual | Trend | $\frac{\text{Actual}}{\text{Trend}}$ |
|---|---|---|---|---|
| 1964 | 1 | 5.1 | 5.49 | .93 |
| | 2 | 6.1 | 5.70 | 1.07 |
| | 3 | 5.7 | 5.94 | .96 |
| | 4 | 6.3 | 6.21 | 1.01 |
| 1965 | 1 | 6.2 | 6.46 | .96 |
| | 2 | 7.2 | 6.73 | 1.07 |
| | 3 | 6.6 | 7.00 | .94 |
| | 4 | 7.5 | 7.28 | 1.03 |
| 1966 | 1 | 7.2 | 7.53 | .96 |
| | 2 | 8.4 | 7.68 | 1.09 |
| | 3 | 7.4 | | |
| | 4 | 7.9 | | |

giving seasonal indexes of

.94, 1.08, .95, and 1.03

**Problem 9.18.** The links are

| Link | 1963 | 1964 | 1965 | 1966 | Average |
|---|---|---|---|---|---|
| 2nd/1st | 130% | 120% | 116% | 117% | 120.8% |
| 3rd/2nd | 92 | 93 | 92 | 88 | 91.3 |
| 4th/3rd | 115 | 111 | 114 | 107 | 111.8 |
| 1st/4th | — | 93 | 98 | 96 | 95.7 |

$$l_1 \times l_2 \times l_3 \times l_4 = 1.18 = 1 + k$$

$$1 - \frac{k}{4} = .955$$

| Quarter | Link | Adj. Links | Accumulated Links | Seasonal Index |
|---|---|---|---|---|
| 1 | | | 100.0% | .94 |
| 2 | 120.8 | 115.4 | 115.4 | 1.09 |
| 3 | 91.3 | 87.2 | 100.6 | .95 |
| 4 | 111.8 | 106.8 | 107.4 | 1.02 |
| | | | 423.4 | 4.00 |

**Problem 9.19.**

| Year | Quarter | Trend | Seasonal Index | Seasonally Adj. Trend | Actual | Difference |
|---|---|---|---|---|---|---|
| 1963 | 1 | 4.49 | .94 | 4.22 | 4.0 | +.22 |
| | 2 | 4.73 | 1.09 | 5.16 | 5.2 | −.04 |
| | 3 | 4.98 | .95 | 4.73 | 4.8 | −.07 |
| | 4 | 5.22 | 1.02 | 5.32 | 5.5 | −.18 |
| 1964 | 1 | 5.47 | .94 | 5.14 | 5.1 | +.04 |
| | 2 | 5.71 | 1.09 | 6.22 | 6.1 | +.12 |
| | 3 | 5.95 | .95 | 5.65 | 5.7 | −.05 |
| | 4 | 6.20 | 1.02 | 6.32 | 6.3 | +.02 |
| 1965 | 1 | 6.44 | .94 | 6.05 | 6.2 | −.15 |
| | 2 | 6.69 | 1.09 | 7.29 | 7.2 | +.09 |
| | 3 | 6.93 | .95 | 6.58 | 6.6 | −.02 |
| | 4 | 7.17 | 1.02 | 7.31 | 7.5 | −.19 |
| 1966 | 1 | 7.42 | .94 | 6.97 | 7.2 | −.23 |
| | 2 | 7.66 | 1.09 | 8.35 | 8.4 | −.05 |
| | 3 | 7.91 | .95 | 7.51 | 7.4 | +.11 |
| | 4 | 8.15 | 1.02 | 8.31 | 7.9 | +.41 |

The last quarter of 1966 shows the greatest variation from the seasonally adjusted trend. However, at the ends of any table, a straight line trend may not be too accurate.

**Problem 9.20.**

| Month | Production | Working Days | Production per Working Day |
|---|---|---|---|
| 1 | 8540 | 22 | 388 |
| 2 | 7415 | 19 | 390 |
| 3 | 8265 | 21 | 394 |
| 4 | 8274 | 21 | 394 |
| 5 | 9210 | 23 | 400 |
| 6 | 8107 | 20 | 405 |

The sixth month was most productive per working day.

# 10

# Index Numbers, Scores and Rates

**10.1. Introduction.** An important part of statistical work is the reduction of large volumes of data to forms in which comparisons can be made and deductions drawn. Thus, the mean, standard deviation, and other functions describe a frequency distribution; the trend and seasonal indexes describe a time series.

*Index Numbers* are statistical measures of groups of related data and are used to compare such data over time, over territory, and in other ways.

*Scores* are used to compare data, often grouped data, for individuals with the whole class of similar individuals.

*Rates* are used in vital statistics to provide an index for a non-homogeneous group of data.

**10.2. Index Numbers.** Index numbers are indicative of various aspects of industry and commerce—cost of living, unemployment, production, wages, etc. They enable us to compare readily, items such as these over periods of time and space. Thus, we have index numbers of the cost of food. Such numbers will vary with the date and also with the area of the country to which they refer. Index numbers normally start with a base of 100 at a particular time for the whole country. It will be seen that index numbers provide time series, discussed in the last chapter, and are subject to analysis as to trend, seasonal movement, etc. Index numbers are often calculated also by territory so that the relative amount of unemployment in different states, for example, may be compared.

**10.3. Problems of Constructing Index Numbers.** Practically every index number presents special problems which are peculiar to itself, but we can illustrate the nature of these problems by considering one example, the *Consumer Price Index*. Clearly a large number of factors enter into the cost of living: housing, food, transportation, etc. A *weighted average* must be used to represent these costs based on periodic sample studies. If the cost of food goes up 10%, but other costs remain the same, the cost of living goes up by that proportion of 10% which the cost of food bears to the total cost of living. This proportion will vary from family to family, and an average proportion must be used. It will also vary with time and territory and it is here that the major difficulty arises in designing a suitable index number. Further, when the cost of food rises, not all foods go up in price by the same amount. A typical "*market basket*" has to be designed to be representative of the average household's purchase of food. Such a basket will, however, vary with time and if some particular food becomes very expensive, the public will shift to other cheaper alternatives.

EXAMPLE 10.1. An index is to be designed to provide a relative measure of the cost of the hobby of photography. It is suggested that this should consist of (1) the cost of an average camera plus (2) the cost of a roll of film. What would be wrong with such a measure?

*Solution.* First, in photography the largest portion of the cost of the hobby is that of processing films and this has been entirely ignored in the measure. Second, many films are bought during the lifetime of a camera. The index should be in the form

$$w_1 \times c_1 + w_2 \times c_2 + w_3 \times c_3$$

where $c_1$, $c_2$, and $c_3$ are the cost of cameras, film, and processing; and $w_1$, $w_2$, and $w_3$ are appropriate weights.

**10.4. Theoretical Considerations.** The ideal index number should meet certain theoretical tests.

1. *Time Reversal Test.* The base date used for an index number should not affect the index. Thus, if calculations are made for the same index with base dates $t_0$ and $t_1$, we have

| Actual Date | Base Date $t_0$ | Base Date $t_1$ |
|---|---|---|
| | Index Number | |
| $t_0$ | 100 | $N'$ |
| $t_1$ | $N$ | 100 |

Then, for consistency, the ratios

$$\frac{100}{N} \quad \text{and} \quad \frac{N'}{100}$$

should be the same.

2. *Factor Reversal Test.* If indexes are constructed of price, quantity, and total value ($P_t$, $Q_t$, and $V_t$) respectively, then for any time $t$

$$\frac{P_t Q_t}{P_0 Q_0} \quad \text{should equal} \quad \frac{V_t}{V_0}$$

In practice, most indexes come close to meeting these tests but do not meet them exactly.

**10.5. Index Weights.** For clarity, the following discussion will refer to a price index. The remarks apply equally to other indexes with suitable changes of wording. A *price index* consists of a number of individual prices $p_t$, where $t$ indicates the time at which the prices are determined. With these prices are associated weights $w$. If the index at time 0 is 100, the expression

$$100 \times \frac{\Sigma\, wp_n}{\Sigma\, wp_0} \tag{10.1}$$

can be used as the index at time $n$. If only one item is involved it will be seen that this reduces to

$$\frac{100\, p_n}{p_0} \tag{10.2}$$

If at time $n$ all items are $(1 + k)$ times the price in year 0, both Formulas 10.1 and 10.2 become

$$100\,(1 + k)$$

One of the most common methods of selecting the weights is to make them proportional to the quantities purchased ($q_0$) in the base year. This is called the *base year* or *Laspeyres' index* and may be written

$$\frac{\Sigma\,(q_0 p_n)}{\Sigma\,(q_0 p_0)}$$

Another method is to make the weights proportional to the quantities purchased ($q_n$) in the year $n$. This is the *given year* or *Paasche's index*,

$$\frac{\Sigma\,(q_n p_n)}{\Sigma\,(q_n p_0)}$$

Two indexes which are more satisfactory than the above, but rather more complicated are the

$$\textit{Marshall-Edgeworth index} = \frac{\Sigma\,(q_0 + q_n)\,p_n}{\Sigma\,(q_0 + q_n)\,p_0}$$

and

$$\textit{Fisher's Ideal price index} = \sqrt{\left(\frac{\Sigma\, q_0 p_n}{\Sigma\, q_0 p_0}\right)\left(\frac{\Sigma\, q_n p_n}{\Sigma\, q_n p_0}\right)}$$

The latter satisfies the time reversal and factor reversal tests.

EXAMPLE 10.2. An index is based on two commodities ($A$ and $B$) only. The prices and quantities sold in the base year (year 0) and a later year (year $n$) are

| **Year** | **Price (p)** | | **Quantity (q)** | |
|---|---|---|---|---|
| | **A** | **B** | **A** | **B** |
| 0 | 150 | 150 | 200 | 100 |
| $n$ | 300 | 150 | 100 | 200 |

If the price index is 100 in year 0, calculate the index in year $n$ for the four indexes described in Section 10.5.
*Solution.*

**Base Year Method (Laspeyres')**

$q_0 \times p_t$

| **Year t** | **A** | **B** | **A + B** | **Index** |
|---|---|---|---|---|
| 0 | 30,000 | 15,000 | 45,000 | 100 |
| $n$ | 60,000 | 15,000 | 75,000 | 167 |

**Given Year Method (Paasche's)**

$q_n \times p_t$

| **Year t** | **A** | **B** | **A + B** | **Index** |
|---|---|---|---|---|
| 0 | 15,000 | 30,000 | 45,000 | 100 |
| $n$ | 30,000 | 30,000 | 60,000 | 133 |

**Marshall-Edgeworth**

$(q_0 + q_n)p_t$

| Year t | A | B | A + B | Index |
|---|---|---|---|---|
| 0 | 45,000 | 45,000 | 90,000 | 100 |
| $n$ | 90,000 | 45,000 | 135,000 | 150 |

**Fisher's Ideal**

With this method, the index is the square root of the multiple of the Base year and Given year methods.

| Year t | Base Year Index | Given Year Index | Fisher's Ideal Index |
|---|---|---|---|
| 0 | 100 | 100 | $\sqrt{100 \times 100} = 100$ |
| $n$ | 167 | 133 | $\sqrt{167 \times 133} = 149$ |

It will be noted that the Base year method develops a higher index, and the Given year method develops a lower index than the Marshall-Edgeworth or the Fisher methods. This is usually the case because quantities bought tend to increase in the comparatively less expensive items. However, the differences are not normally so marked as in this specially selected example.

EXAMPLE 10.3. A cost of food index is to be constructed from the following six items.

| Item | Price 1950 | Price 1960 | Quantity 1950 | Quantity 1960 |
|---|---|---|---|---|
| Bread | 7.0 | 8.1 | 75 | 70 |
| Meat | 50.5 | 57.4 | 23 | 27 |
| Vegetables | 8.4 | 8.9 | 67 | 70 |
| Milk Products | 45.3 | 47.2 | 10 | 15 |
| Beverages | 61.7 | 63.8 | 10 | 12 |

Using 1950 as the base year with index of 100, calculate, by the base year method, the index of 1960.

*Solution.* The calculations can be made as follows.

| | $p_0$ | $q_0$ | $p_0q_0$ | $p_{10}$ | $q_0$ | $p_{10}q_0$ |
|---|---|---|---|---|---|---|
| Bread | 7.0 | 75 | 525.0 | 8.1 | 75 | 607.5 |
| Meat | 50.5 | 23 | 1161.5 | 57.4 | 23 | 1320.2 |
| Vegetables | 8.4 | 67 | 562.8 | 8.9 | 67 | 596.3 |
| Milk Products | 45.3 | 10 | 453.0 | 47.2 | 10 | 472.0 |
| Beverages | 61.7 | 10 | 617.0 | 63.8 | 10 | 638.0 |
| | | | 3319.3 | | | 3634.0 |

$$\text{Index for 1960} = 100 \times \frac{3634.0}{3319.3} = 109.5$$

It is important to realize that in practice most indexes will involve hundreds of items, not just a few as in this example.

EXAMPLE 10.4. Using the data in Example 10.3, calculate, by the given year method, the index for 1960.
*Solution.* The calculations can be made as follows,

| | $p_0$ | $q_{10}$ | $p_0q_{10}$ | $p_{10}$ | $q_{10}$ | $p_{10}q_{10}$ |
|---|---|---|---|---|---|---|
| Bread | 7.0 | 70 | 490.0 | 8.1 | 70 | 567.0 |
| Meat | 50.5 | 27 | 1363.5 | 57.4 | 27 | 1549.8 |
| Vegetables | 8.4 | 70 | 588.0 | 8.9 | 70 | 623.0 |
| Milk Products | 45.3 | 15 | 679.5 | 47.2 | 15 | 708.0 |
| Beverages | 61.7 | 12 | 740.4 | 63.8 | 12 | 765.6 |
| | | | 3861.4 | | | 4213.4 |
| | **Index** | | = **100** | | | = **109.1** |

EXAMPLE 10.5. Using the data in Example 10.3, calculate the Marshall-Edgeworth cost-of-food index for 1960, based on 100 for 1950.
*Solution.* In this index, the weights are the sum of the quantities in the base year and the given year.

| | $p_0$ | $q_0+q_{10}$ | $p_0(q_0 + q_{10})$ | $p_{10}$ | $q_0+q_{10}$ | $p_0(q_0 + q_{10})$ |
|---|---|---|---|---|---|---|
| Bread | 7.0 | 145 | 1015.0 | 8.1 | 145 | 1174.5 |
| Meat | 50.5 | 50 | 2525.0 | 57.4 | 50 | 2870.0 |
| Vegetables | 8.4 | 137 | 1150.8 | 8.9 | 137 | 1219.3 |
| Milk Products | 45.3 | 25 | 1132.5 | 47.2 | 25 | 1180.0 |
| Beverages | 61.7 | 22 | 1357.4 | 63.8 | 22 | 1403.6 |
| | | | 7180.7 | | | 7847.4 |
| | **Index** | | = **100** | | | = **109.3** |

It should be noted that, if the Base and Given year indexes have been calculated, there is no need to go through the complete calculations set out above, since

$$7180.7 = 3319.3 + 3861.4$$

and

$$7847.4 = 3634.0 + 4213.4$$

EXAMPLE 10.6. Using the data in Example 10.3, calculate the Fisher's Ideal index for 1960 based on 100 for 1950.

*Solution.* Fisher's Ideal index is the square root of the Base year index multiplied by the Given year index.

$$\text{Fisher's Ideal index} = \sqrt{109.5 \times 109.1} = 109.3$$

It will be noted that the differences among the four indexes in Examples 10.3 to 10.6 are small, and are within the range of variation which might be expected from other causes, such as judgment used in determining the items entering into the index.

**10.6. Quantity Index Numbers.** If, instead of indexes of prices, indexes of production or consumption are required, the most appropriate weights are often the values (prices), and hence the formula for quantity indexes are the same as for price indexes with $p$ and $q$ interchanged.

**10.7. Use of the Geometric or Harmonic Means.** There are certain theoretical justifications for using the suitably weighted geometric or harmonic means instead of the weighted arithmetic mean in establishing index numbers. These practices are not used widely at present because such indexes are not as easily understood and are more laborious to calculate. They may become more common with the greater use of data processing equipment.

**10.8. Cost of Living Adjustments.** Many long period time series of income, wages, sales, etc., are difficult to interpret because of changes in the cost of living. For this reason, such figures are sometimes divided by the Consumer Price Index, to provide cost of living adjusted statistics.

**10.9. Center of Population.** The center of population of the United States is that point upon which the United States would balance if it were a rigid plane without weight and each individual was assumed to have an equal weight. It provides a simple single index to show the shift of the population over long periods of time.

In 1790, the center was 23 miles to the east of Baltimore, in Maryland. From 1820 to 1850, it was in West Virginia; in 1860 and 1870, it was in Ohio; in 1880, it was just southwest of Cincinnati; from 1890 to 1940, it was in Indiana, and since 1950 it has been in Illinois. The center is moving west at about 4 or 5 miles a year.

**10.10. Scores.** If a single test is taken by a body of students, the *marks* or *scores* provide a measure of the student's abilities. However, when different tests are involved, it is necessary to adjust the results for the difficulty of the tests. The procedure was explained when *standard scores* were discussed in Chapter 4.

*Intelligence tests* are designed so that a student with *normal intelligence* will have an intelligence quotient of 100%. If his *intelligence quotient* differs from 100% his mental age is defined as his true or *chronological age* multiplied by this percentage.

$$\text{I.Q.} = \frac{\text{MA}}{\text{CA}} = \frac{\text{mental age}}{\text{chronological age}} \qquad (10.3)$$

Intelligence tests are broken down into various subjects; arithmetic, reading, etc. Quotients for individual subjects can be developed similarly.

**10.11. Demography.** Demography is the study of human populations by statistical methods. The subject is also referred to as *vital statistics.* In any given population, the ratio of deaths in a year to the total population in the middle of the year is calculated and this is called the death rate. Similarly, there is a *birth rate*, a *marriage rate*, and a *morbidity* (sickness) *rate*. The morbidity rate is the ratio of the average number of *people who are sick* at any time (not the average number of people becoming sick).

These rates, called *crude rates*, are not very useful for serious studies. The reason for this is that the probability of, say, dying varies very greatly with the age. The *age distribution* of the population being studied will greatly affect the crude death data. This is equally true of births, marriages, and sickness. For this reason *standardized rates* are used.

If the population is subdivided according to age (or short groups of ages) and the deaths are similarly subdivided, the ratios which these figures develop will be death rates at various ages. If these death rates are applied to a *standard population* (which may be selected in various ways), a *standardized death rate* is arrived at, which is suitable for properly comparing the mortality of different populations. Standardized rates are weighted averages of the data analyzed by age.

*Fertility rates* are usually calculated separately by quinquennial age groups of the mother; overall fertility rates may be obtained by dividing the number of live births in a year by the number of women aged 15 to 44 years.

A *Net Reproductive Rate* of 1000 means that each generation would just replace itself if birth and death rates specified continued indefinitely and there was no migration. Rates above 1000 imply gaining population, rates below imply declining population.

**10.12. Population Pyramids.** In order to display the distribution of a population by age and sex, a population pyramid is often used as illustrated below

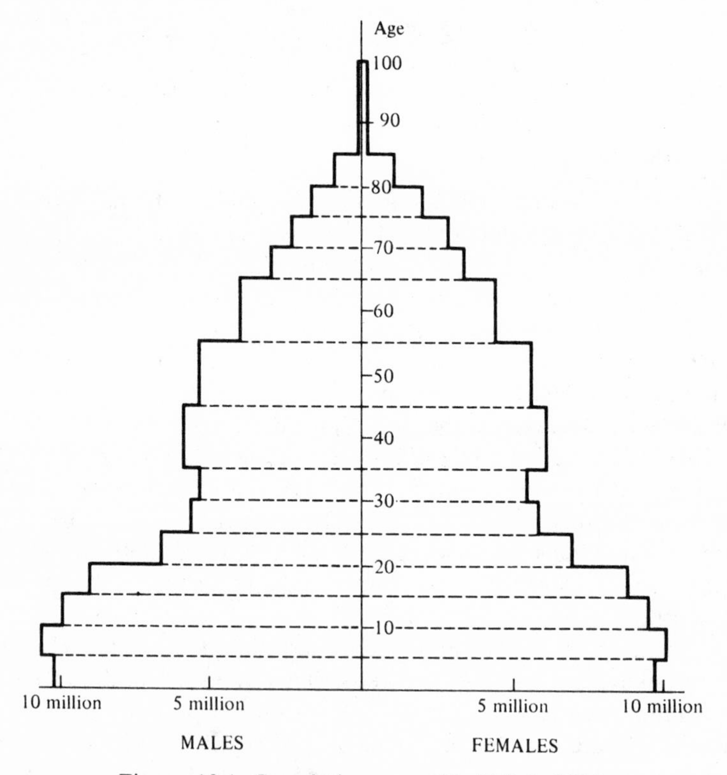


Figure 10.1. Population pyramid. U.S.A. 1966.

**10.13. Expectation of Life.** A useful measure of overall mortality is the expectation of life, which is the average number of years a person can expect to live according to a certain mortality experience. Expectations of life can be calculated at any age from 0 upwards. Since they do not allow for mortality trends, the calculated expectations normally understate the true expectation. Further, they apply an *average* life, including both healthy and unhealthy. According to the United States census data, the expectations of life at birth were as follows.

| Year of Birth | Expectation of Life |
|---|---|
| 1920 | 54.1 |
| 1930 | 59.7 |
| 1940 | 62.9 |
| 1950 | 68.2 |
| 1960 | 69.7 |
| 1965 | 70.2 |

In 1965, the expectation of life at birth was 66.8 for men and 73.7 for women.

## Problems

**Problem 10.1.** Indicate the type of statistics for which index numbers are used by the U.S. Government.

**Problem 10.2.** A certain commodity represents $x\%$ of a price index, and the price of this commodity increases by $y\%$, while all other commodities making up the index remain unchanged. What will be the resulting percentage change in the price index?

**Problem 10.3.** Over a series of years the prices of all the items making up a certain price index remain unchanged, but an expensive item is purchased less and less each year. What will be the effect on

(1) Base Year, or Laspeyres' Index
(2) Given Year, or Paasche's Index?

**Problem 10.4.** A certain index number has the following values.

| 1960 | 1961 | 1962 | 1963 | 1964 | 1965 | 1966 |
|---|---|---|---|---|---|---|
| 105 | 110 | 123 | 128 | 128 | 154 | 165 |

It is decided to introduce a new base, 1966 = 100, and to recalculate the indexes for the previous years. If the index satisfies the time reversal test, what are the revised values?

**Problem 10.5.** An index is made up of three items. The weights in year 0 are: *A* 30%, *B* 20%, *C* 50%.

The prices of the items in year 0 and year *n* are:

| Year | 0 | n |
|---|---|---|
| *A* | 54 | 63 |
| *B* | 42 | 42 |
| *C* | 49 | 53 |

If the index for year 0 is 100, what is the index for year *n*, using the base year (Laspeyres') method?

**Problem 10.6.** Use the data of Problem 10.5, and assume weights of *A* 20%, *B* 20%, and *C* 60% in the year *n*. What is the index in year *n*, corresponding to 100 in year 0, using the given year (Paasche's) method?

**Problem 10.7.** Using the data of Problems 10.5 and 10.6, what is the Marshall-Edgeworth index?

**Problem 10.8.** Show mathematically that Fisher's Ideal price index satisfies the time reversal test.

**Problem 10.9.** Show mathematically that Fisher's Ideal price index satisfies the factor reversal test.

**Problem 10.10.** The following table gives the U.S. Gross National Product in billions of dollars for various years and also the Consumer Price Index for the same years. Adjust the G.N.P. to a level Consumer Price, so as to remove the element of inflation for the G.N.P.

| **Year** | 1940 | 1945 | 1950 | 1955 | 1960 | 1965 |
|---|---|---|---|---|---|---|
| **G.N.P.** | 99.7 | 211.9 | 284.8 | 398.0 | 503.7 | 683.9 |
| **C.P.I.** | 48.8 | 62.7 | 83.8 | 93.3 | 103.1 | 109.9 |

**Problem 10.11.** Two cities are 100 miles apart. City *A* has a population of two million and city *B* has a population of three million. What is the location of the center of population of the two cities combined?

**Problem 10.12.** A child aged 8 has an I.Q. of 125. What is his mental age?

**Problem 10.13.** A statistician goes through the notice of deaths in his local paper each day and analyzes the figures by (1) occupation and (2) age at death. By taking the average age of death over a number of years, he establishes which are the most healthy occupations. List four reasons why his results might not be sound.

**Problem 10.14.** The *rate of mortality* is defined as the number of people expected to die between age $x$ and $x + 1$ out of a given number of people exactly aged $x$. The *death rate* at age $x$ is the number of people dying in a year at age $x$, expressed as a ratio to the number of people living at age $x$ in the middle of the year. Find a relationship between these two rates.

**Problem 10.15.** Assume the death rates at age 60, 65, and 70 are .03, .04, and .06. What is the crude death rate for a population ($A$) consisting of 1000 persons aged 60, 500 persons aged 65, and 100 persons aged 70, if they experience the assumed rates? What is the death rate for a population ($B$) consisting of 500 persons aged 60, 500 aged 65, and 500 aged 70?

**Problem 10.16.** In the above problem, what would be the standardized death rate if the standard population were as follows?

| Age | Population |
|---|---|
| 60 | 65,000 |
| 65 | 55,000 |
| 70 | 45,000 |

**Problem 10.17.** What increments, other than death, cause the population of a territory to change from year to year?

**Problem 10.18.** What would be the effect on a population pyramid of

(1) A high rate of immigration
(2) A declining birthrate?

Assume these features have continued for a long time.

**Problem 10.19.** The expectation of life at age $x$ is the average number of years a person can expect to live after age $x$. The function used to represent the expectation of life at age $x$ is

$$\overset{\circ}{e}_x$$

(1) What will be the average age at death of persons now aged $x$ according to the mortality assumed?

(2) Does the expectation of life decrease with age?

**Problem 10.20.** What evidence is there that women live longer than men?

## Solutions

**Problem 10.1.** Indexes are used extensively for *Prices* (Wholesale, Consumer, Retail, etc.) and also for *Production* (both as a measure of *Quantity* and of *Value*).

**Problem 10.2.**

$$\frac{x}{100} \times y\%$$

**Problem 10.3.** Since no prices have changed over the years, neither index will have changed.

**Problem 10.4.**

| 1960 | 1961 | 1962 | 1963 | 1964 | 1965 | 1966 |
|---|---|---|---|---|---|---|
| 64 | 67 | 75 | 78 | 78 | 93 | 100 |

**Problem 10.5.**

$$\frac{(.3 \times 63) + (.2 \times 42) + (.5 \times 53)}{(.3 \times 54) + (.2 \times 42) + (.5 \times 49)} = 110$$

**Problem 10.6.** 109
**Problem 10.7.** 109
**Problem 10.8.** Assume the prices and quantities of the various items entering into the index are:

| Time 0 | | Time n | |
|---|---|---|---|
| **Price** | **Quantity** | **Price** | **Quantity** |
| $p_0$ | $q_0$ | $p_n$ | $q_n$ |

Taking 100 as the index at time 0, Fisher's index at time $n$ is

$$\sqrt{\left(\frac{\Sigma\, q_0 p_n}{\Sigma\, q_0 p_0}\right)\left(\frac{\Sigma\, q_n p_n}{\Sigma\, q_n p_0}\right)}$$

Taking 100 as the index at time $n$, Fisher's index at time 0 is obtained by interchanging the suffixes 0 and $n$.

$$\sqrt{\left(\frac{\Sigma\, q_n p_0}{\Sigma\, q_n p_n}\right)\left(\frac{\Sigma\, q_0 p_0}{\Sigma\, q_0 p_n}\right)}$$

We see at once that the second expression is the reciprocal of the first, proving time reversal.

**Problem 10.9.** Proceed as in the previous solution. For the factor reversal test interchange $p$ and $q$ giving

$$\sqrt{\left(\frac{\Sigma\, p_0 q_n}{\Sigma\, p_0 q_0}\right)\left(\frac{\Sigma\, p_n q_n}{\Sigma p_n q_0}\right)}$$

Multiplying this expression by the normal Fisher formula we have,

$$\frac{\Sigma p_n q_n}{\Sigma p_0 q_0} = \frac{\text{Total Value at time } n}{\text{Total Value at time } 0}$$

**Problem 10.10.** The G.N.P. should be divided by the C.P.I. to produce the adjusted G.N.P. These figures can then be best expressed using an index of 100 for 1940.

| **Year** | **1940** | **1945** | **1950** | **1955** | **1960** | **1965** |
|---|---|---|---|---|---|---|
| **GNP / CPI** | 204.3 | 338.0 | 339.9 | 426.6 | 488.6 | 622.3 |
| **Adjusted to 100 for 1940** | 100 | 165 | 166 | 209 | 239 | 305 |

**Problem 10.11.** We must assume that the center of population of each city is at the center of the city. The center of population of the two cities combined will, by symmetry, lie on the line joining the two cities. Let it be $x$ miles from city $A$. Then,

$$(2{,}000{,}000 \times 0) + (3{,}000{,}000 \times 100) = 5{,}000{,}000 \times x$$

$$\text{or} \qquad x = \frac{300}{5} = 60 \text{ miles}$$

The center of population is 60 miles from city $A$ and 40 miles from city $B$.

**Problem 10.12.**

$$\begin{aligned}\text{Mental Age} &= \text{I.Q.} \times (\text{chronological age}) \\ &= 1.25 \times 8 \\ &= 10 \text{ years}\end{aligned}$$

**Problem 10.13.**

(1) *Changes in types of occupation available over the years.* For example, the list would be unlikely to include many old electronic data programmers or any young steam engine drivers.

(2) *Date of entry into occupation classification.* A person has to live quite a time before he becomes a bishop or an orchestra conductor, so that these and similar occupations will show old average ages at death.

(3) *Migration.* People in the higher paid occupations are likely to move to resort type areas on retirement and may not be listed on death.

(4) *Change in occupation class.* Some occupations, such as professional football players, cannot be continued throughout life and all deaths recorded for these classes will be young.

**Problem 10.14.** Consider 100,000 people all assumed to be exact age $x$ on January 1st of a certain year. Let $q_n$ be the rate of mortality. Then 100,000 $q_x$ will die in the year and approximately one-half of these will die in the first six months. Hence, the number of people aged $x$ in the middle of the year is approximately

$$100{,}000\left(1 - \frac{1}{2}q_x\right)$$

and the death rate is

$$\frac{100{,}000\, q_x}{100{,}000\,(1 - {}^1\!/_2\, q_x)}$$

or

$$\frac{2q_x}{2 - q_x}$$

**Problem 10.15.**

| Age | Death Rate | A Number | A Deaths | B Number | B Deaths |
|---|---|---|---|---|---|
| 60 | .03 | 1000 | 30 | 500 | 15 |
| 65 | .04 | 500 | 20 | 500 | 20 |
| 70 | .06 | 100 | 6 | 500 | 30 |
| | | 1600 | 56 | 1500 | 65 |
| | **Crude Death Rate =** | | **.035** | | **.043** |

**Problem 10.16.**

| Standardized Population | Death Rate | Deaths |
|---|---|---|
| 65,000 | .03 | 1,950 |
| 55,000 | .04 | 2,200 |
| 45,000 | .06 | 2,700 |
| 165,000 | | 6,850 |

**Standardized Death Rate = .042**

**Problem 10.17.** Births and migration.

**Problem 10.18.**

(1) Immigrants normally are heavily weighted in the age 20–30 group, and in men rather than women. Hence, the pyramid

will have a similar base (fewer children) relative to its size at the older ages.

(2) A declining birthrate will also cause a smaller base relative to the higher ages.

**Problem 10.19.**

(1) $\mathring{e}_x + x$

(2) Normally, the expectation of life decreases as the age increases, once the first year or so of age is passed. The following figures are from U.S. mortality for the year 1965.

| Age | Expectation of Life |
|---|---|
| 0 | 70.2 |
| 1 | 70.9 |
| 2 | 70.0 |
| 10 | 62.3 |
| 20 | 52.7 |
| 50 | 25.5 |

**Problem 10.20.** There is ample statistical evidence that women live longer than men. The expectation of life of women at age 0 is about 7 years longer than men (see figure in text). Further, the very high proportion of women to men at the older ages proves the point.

# 11

# Correlation and Regression

**11.1. Introduction.** Hitherto we have considered only a single variable and the frequencies, distribution, and analysis thereof. In this chapter, we shall consider the analysis of two variables $x$ and $y$, and the frequencies with which *pairs of values* occur. This is called *bivariate analysis.*

**11.2. Scatter Diagrams.** Thirty students sit for two tests in the same subject a month apart. Their scores are

| First Test | Second Test | First Test | Second Test |
|---|---|---|---|
| 87 | 80 | 72 | 70 |
| 80 | 95 | 70 | 63 |
| 32 | 32 | 62 | 65 |
| 65 | 72 | 85 | 90 |
| 55 | 67 | 55 | 47 |
| 30 | 47 | 45 | 40 |
| 82 | 82 | 68 | 70 |
| 90 | 85 | 77 | 73 |
| 38 | 45 | 68 | 55 |
| 72 | 80 | 67 | 65 |
| 60 | 70 | 62 | 58 |
| 45 | 50 | 52 | 60 |
| 30 | 30 | 57 | 60 |
| 90 | 93 | 77 | 85 |
| 73 | 75 | 80 | 78 |

We often desire to examine lists such as these to see what are the relations between the two lists. In this case we note, as we should expect since the tests are in the same subject, that large scores in the first test correspond to large scores in the second test. This is called *positive correlation.* If large scores in one test had corresponded to small scores in the second test this would be *negative correlation.* One way of displaying these results is a *scatter diagram* obtained by plotting the two scores on graph paper.

EXAMPLE 11.1. Plot a scatter diagram for the data given above.
*Solution.* Each dot represents an individual pair of scores and the results are shown in Figure 11.1.

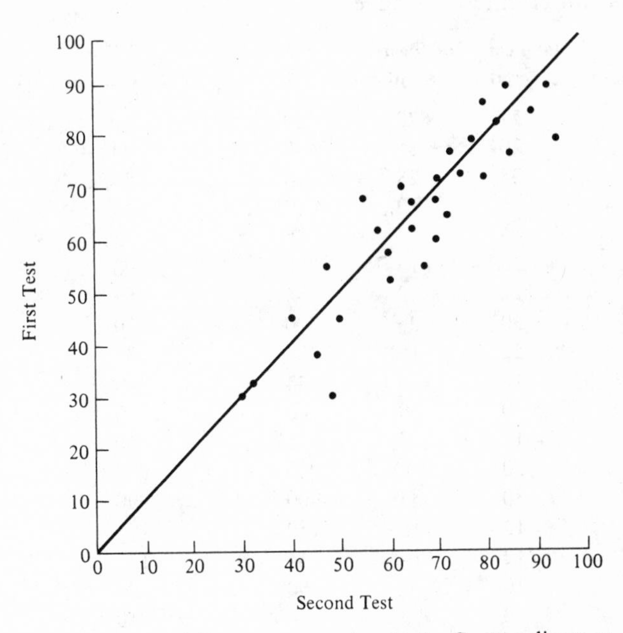


Figure 11.1. Students scores on two tests. Scatter diagram.

We note that the results are positively correlated. When the points are dotted in a haphazard way on the graph, the two functions have *no correlation.*

**11.3. Lines of Regression.** In Chapter 8 it was shown how the method of least squares could be used to find an approximate value of $y$ in terms of $x$. Such a fitted curve is called the *line of regression of y on x.*

If the functions $x$ and $y$ are interchanged, we obtain the line of regression of $x$ *on* $y$. These two lines are not normally the same. If $y$ is related to $x$ by a formula, a single line through all the points will be both the line of regression of $y$ on $x$ and of $x$ on $y$. If the curves fitted are straight lines, this is referred to as *linear regression.* The two lines will meet in a point $\bar{x}$, $\bar{y}$, the center of gravity (or centroid) of the data.

EXAMPLE 11.2. Calculate the lines of regression for the data given in Section 11.2.

*Solution.* Proceeding as in Chapter 8 and using origin of 60 marks for each test, we have

| Student | 1st Test $x - 60$ | 2nd Test $y - 60$ | $(x - 60)^2$ | $(x - 60)(y - 60)$ | $(y - 60)^2$ |
|---|---|---|---|---|---|
| 1 | 27 | 20 | 729 | 540 | 400 |
| 2 | 20 | 35 | 400 | 700 | 1,225 |
| 3 | −28 | −28 | 784 | 784 | 784 |
| 4 | 5 | 12 | 25 | 60 | 144 |
| 5 | −5 | 7 | 25 | −35 | 49 |
| 6 | −30 | −13 | 900 | 390 | 169 |
| 7 | 22 | 22 | 484 | 484 | 484 |
| 8 | 30 | 25 | 900 | 750 | 625 |
| 9 | −22 | −15 | 484 | 330 | 225 |
| 10 | 12 | 20 | 144 | 240 | 400 |
| 11 | 0 | 10 | 0 | 0 | 100 |
| 12 | −15 | −10 | 225 | 150 | 100 |
| 13 | −30 | −30 | 900 | 900 | 900 |
| 14 | 30 | 33 | 900 | 990 | 1,089 |
| 15 | 13 | 15 | 169 | 195 | 225 |
| 16 | 12 | 10 | 144 | 120 | 100 |
| 17 | 10 | 3 | 100 | 30 | 9 |
| 18 | 2 | 5 | 4 | 10 | 25 |
| 19 | 25 | 30 | 625 | 750 | 900 |
| 20 | −5 | −13 | 25 | 65 | 169 |
| 21 | −15 | −20 | 225 | 300 | 400 |
| 22 | 8 | 10 | 64 | 80 | 100 |
| 23 | 17 | 13 | 289 | 221 | 169 |
| 24 | 8 | −5 | 64 | −40 | 25 |
| 25 | 7 | 5 | 49 | 35 | 25 |
| 26 | 2 | −2 | 4 | −4 | 4 |
| 27 | −8 | 0 | 64 | 0 | 0 |
| 28 | −3 | 0 | 9 | 0 | 0 |
| 29 | 17 | 25 | 289 | 425 | 625 |
| 30 | 20 | 18 | 400 | 360 | 324 |
| | 287 | 318 | 9,424 | 8,909 | 9,794 |
| | −161 | −136 | | −79 | |
| | 126 | 182 | | 8,830 | |

The normal equations for the line of regression of $y - 60$ on $x - 60$ are

$$182 = 30a_0 + 126a_1$$
$$8{,}830 = 126a_0 + 9{,}424a_1$$

Giving $$a_0 = 2.24 \qquad a_1 = .91$$
$$(y - 60) = 2.24 + .91(x - 60)$$

The normal equations of the line or regression of $x - 60$ on $y - 60$ are

$$126 = 30b_0 + 182b_1$$
$$8{,}830 = 182b_0 + 9{,}794b_1$$

Giving $$(x - 60) = -1.44 + .93(y - 60)$$

The two lines of regression cross at the centroid (the center of gravity) of the data and using this point as origin the lines are

$$\bar{y} = .91\bar{x}$$
$$\bar{x} = .93\bar{y}$$

The centroid is

$$x = 60 + \frac{\Sigma x}{n} = 60 + \frac{126}{30} = 64.2$$

$$y = 60 + \frac{\Sigma y}{n} = 60 + \frac{182}{30} = 66.1$$

**11.4. Tabulation of Grouped Data.** When the volume of data is large, a scatter diagram is not practical and a scatter or *distribution table* must be used. In such a table, the data are grouped in class intervals in a similar way to that used earlier in this book, but since we have two variables, the table is *two dimensional*. The following is a *distribution table* for the data in Section 11.2.

DISTRIBUTION TABLE

**Score 1st Test**

| Score 2nd Test | 30– | 40– | 50– | 60– | 70– | 80– | 90– |
|---|---|---|---|---|---|---|---|
| 90– | | | | | | **2** | **1** |
| 80– | | | | | **2** | **2** | **1** |
| 70– | | | | **3** | **3** | **1** | |
| 60– | | | **3** | **2** | **1** | | |
| 50– | | **1** | | **2** | | | |
| 40– | **2** | **1** | **1** | | | | |
| 30– | **2** | | | | | | |

**11.5. Coefficient of Correlation.** Let the equation of the lines of regression (referred to the centroid as origin) be

$$\bar{y} = k_1 \bar{x}$$

and

$$\bar{x} = k_2 \bar{y}$$

If they are identical and correlation between the two functions is complete then

$$\frac{1}{k_2} = k_1$$

or

$$k_1 k_2 = 1$$

The *coefficient of correlation* is defined as

$$r = \sqrt{k_1 k_2}$$

If values of $x$ and $y$ are measured from the centroid,

$$r = \frac{\Sigma xy}{\sqrt{(\Sigma x^2)(\Sigma y^2)}} \tag{11.1}$$

This and Formula 11.2 are called *product-moment formulas.* If values are measured from some other origin,

$$r = \frac{n\Sigma xy - \Sigma x \Sigma y}{\sqrt{[n(\Sigma x^2) - (\Sigma x)^2][n(\Sigma y^2) - (\Sigma y)^2]}} \tag{11.2}$$

This formula may be written

$$r = \frac{p}{\sigma_x \sigma_y} \tag{11.3}$$

where

$$p = \frac{\Sigma xy}{n} - \frac{\Sigma x}{n} \cdot \frac{\Sigma y}{n}$$

$$\sigma_x = \frac{\Sigma x^2}{n} - \left(\frac{\Sigma x}{n}\right)^2$$

$$\sigma_y = \frac{\Sigma y^2}{n} - \left(\frac{\Sigma y}{n}\right)^2$$

EXAMPLE 11.3. Apply Formula 11.3 to obtain the coefficient of correlation of the data in Section 11.2.
*Solution.* In the solution to Example 11.2, $x = 60$, $y = 60$, was

used as an origin. From the figures in that solution

$$\bar{x} = \frac{126}{30} = 4.2$$

$$\bar{y} = \frac{182}{30} = 6.1$$

$$\sigma_x^2 = \frac{9424}{30} - (4.2)^2 = 296.5 \qquad \text{and } \sigma_x = 17.2$$

$$\sigma_y^2 = \frac{9794}{30} - (6.1)^2 = 289.3 \qquad \text{and } \sigma_y = 17.0$$

$$p = \frac{8830}{30} - (4.2)(6.1) = 268.7$$

$$r = \frac{268.7}{(17.2)(17.0)} = .92$$

**11.6. Coefficient of Correlation—Grouped Data.** The methods of handling grouped data in the calculation of the mean and standard deviation have been explained earlier in the book. Completely similar procedures apply to the calculation of the coefficient of correlation. We may proceed to make the calculation for each group in a single tabulation as in Example 11.4, or, as is more usual, set the working-out in a two dimensional table as in Example 11.5 and 11.6.

EXAMPLE 11.4. Calculate the coefficient of correlation for the grouped data in Section 11.4, using a single column tabulation.
*Solution.* Using origin of $x = 64^1/_2$, $y = 64^1/_2$, and 10 marks as class scale, and writing $x = 64^1/_2 + 10u$ and $y = 64^1/_2 + 10v$

| 1st Test | 2nd Test | u | v | Frequency f | fu | fv | fu² | fuv | fv² |
|---|---|---|---|---|---|---|---|---|---|
| 30– | 30– | −3 | −3 | 2 | −6 | −6 | 18 | 18 | 18 |
| 30– | 40– | −3 | −2 | 2 | −6 | −4 | 18 | 12 | 8 |
| 40– | 40– | −2 | −2 | 1 | −2 | −2 | 4 | 4 | 4 |
| 40– | 50– | −2 | −1 | 1 | −2 | −1 | 4 | 2 | 1 |
| 50– | 40– | −1 | −2 | 1 | −1 | −2 | 1 | 2 | 4 |
| 50– | 50– | −1 | −1 | 0 | 0 | 0 | 0 | 0 | 0 |
| 50– | 60– | −1 | 0 | 3 | −3 | 0 | 3 | 0 | 0 |
| 60– | 50– | 0 | −1 | 2 | 0 | −2 | 0 | 0 | 2 |
| 60– | 60– | 0 | 0 | 2 | 0 | 0 | 0 | 0 | 0 |
| 60– | 70– | 0 | 1 | 3 | 0 | 3 | 0 | 0 | 3 |

| 1st Test | 2nd Test | u | v | Frequency f | fu | fv | fu² | fuv | fv² |
|---|---|---|---|---|---|---|---|---|---|
| 70– | 60– | 1 | 0 | 1 | 1 | 0 | 1 | 0 | 0 |
| 70– | 70– | 1 | 1 | 3 | 3 | 3 | 3 | 3 | 3 |
| 70– | 80– | 1 | 2 | 2 | 2 | 4 | 2 | 4 | 8 |
| 80– | 70– | 2 | 1 | 1 | 2 | 1 | 4 | 2 | 1 |
| 80– | 80– | 2 | 2 | 2 | 4 | 4 | 8 | 8 | 8 |
| 80– | 90– | 2 | 3 | 2 | 4 | 6 | 8 | 12 | 18 |
| 90– | 80– | 3 | 2 | 1 | 3 | 2 | 9 | 6 | 4 |
| 90– | 90– | 3 | 3 | 1 | 3 | 3 | 9 | 9 | 9 |
| | | | | 30 | 2 | 9 | 92 | 82 | 91 |

$$\bar{u} = \frac{2}{30} = .067$$

$$\bar{v} = \frac{9}{30} = .300$$

$$\sigma_x^2 = \frac{92}{30} - (.067)^2 = 3.07 - .00 = 3.07$$

$$\sigma_y^2 = \frac{91}{30} - (.300)^2 = 3.03 - .09 = 2.94$$

$$p = \frac{82}{30} - (.067)(.300) = 2.73 - .02 = 2.71$$

$$r = \frac{p}{\sigma_x \sigma_y} = \frac{2.71}{1.75 \times 1.71} = \frac{2.71}{2.99} = .91$$

*It should be particularly noted that at no time during the calculations do we return to the true origin or the true units.*

EXAMPLE 11.5. Calculate the coefficient of correlation for the grouped data in Section 11.4, using a two dimensional table.
*Solution.* The two dimensional tabulation proceeds as follows. Comparison should be made with the previous solution. In the main box, the frequency in the various groupings of $u$ and $v$ from the table in Section 11.4 are shown in the left top corner and the factor $uv$ in the right bottom corner. This is done to facilitate

| v \ u | | 30– | 40– | 50– | 60– | 70– | 80– | 90– | f | fv | $fv^2$ | fuv |
|---|---|---|---|---|---|---|---|---|---|---|---|---|
| | | −3 | −2 | −1 | 0 | 1 | 2 | 3 | | | | |
| 90– | 3 | | | | | | 2 / 6 | 1 / 9 | 3 | 9 | 27 | 21 |
| 80– | 2 | | | | | 2 / 2 | 2 / 4 | 1 / 6 | 5 | 10 | 20 | 18 |
| 70– | 1 | | | | 3 / 0 | 3 / 1 | 1 / 2 | | 7 | 7 | 7 | 5 |
| 60– | 0 | | | 3 / 0 | 2 / 0 | 1 / 0 | | | 6 | 0 | 0 | 0 |
| 50– | −1 | | 1 / 2 | | 2 / 0 | | | | 3 | −3 | 3 | 2 |
| 40– | −2 | 2 / 6 | 1 / 4 | 1 / 2 | | | | | 4 | −8 | 16 | 18 |
| 30– | −3 | 2 / 9 | | | | | | | 2 | −6 | 18 | 18 |
| f | | 4 | 2 | 4 | 7 | 6 | 5 | 2 | 30 | 9 | 91 | 82 |
| fu | | −12 | −4 | −4 | 0 | 6 | 10 | 6 | 2 | | | |
| $fu^2$ | | 36 | 8 | 4 | 0 | 6 | 20 | 18 | 92 | | | |

the calculation of the final column. Some texts show $fuv$ in the right hand corner, others include three figures in each box.

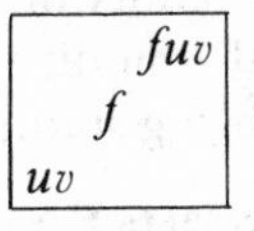


In this case, the top row would read

| | |
|---|---|
| 12 | 9 |
| 2 | 1 |
| 6 | 9 |

From the table we can read off, $n = 30$, $fu = 2$, $fu^2 = 92$, $fv = 9$, $fv^2 = 91$, $fuv = 82$, and the calculation of the coefficient of correlation proceeds as in the previous example.

EXAMPLE 11.6. The following table gives the analysis of group medical insurance plans in a certain study according to (1) size of plan and (2) loss ratio experience.

**Medical Experience Group Plans**

| **Ratio of Actual to Expected Losses** | -50% | 50–119% | 120–199% | 200%- |
|---|---|---|---|---|
| **Size of Plan** | | **Number of Plans** | | |
| Under 50 lives | 630 | 694 | 368 | 309 |
| 50–99 | 285 | 638 | 294 | 77 |
| 100–499 | 113 | 860 | 307 | 44 |
| 500– | 3 | 232 | 70 | 3 |

Assuming the midpoints of the groups are as follows, calculate the coefficient of correlation between size of group and loss experience.

| **Size of Plan** | **Midpoint** |
|---|---|
| Under 50 lives | 30 |
| 50–99 | 75 |
| 100–499 | 200 |
| 500– | 1,000 |

| **Ratio of Actual to Expected** | |
|---|---|
| -50% | 30% |
| 50%–119% | 80% |
| 120%–199% | 150% |
| 200%- | 250% |

*Solution.* We will use as origin the midpoint of the second class interval for each variable. Units of increment will be 100 employees and 100% so that the modified variables applying to the two attributes of the plans being studied are

**Ratio of Actual to Expected Losses**

| **Class Interval** | **Midpoint** | **u** |
|---|---|---|
| -50% | 30% | −.5 |
| 50%–119% | 80% | 0 |
| 120%–199% | 150% | .7 |
| 200%- | 250% | 1.7 |

**Size of Plan**

| **Class Interval** | **Midpoint** | **v** |
|---|---|---|
| Under 50 lives | 30 | −.45 |
| 50–99 | 75 | 0 |
| 100–499 | 200 | 1.25 |
| 500– | 1,000 | 9.25 |

The calculations proceed as follows.

| v \ u | $-.5$ | $0$ | $.7$ | $1.7$ | f | fv | fv² | fuv |
|---|---|---|---|---|---|---|---|---|
| $-.45$ | 630 / .225 | 694 / 0 | 368 / $-.315$ | 309 / $-.765$ | 2001 | $-900$ | 405 | $-211$ |
| $0$ | 285 / 0 | 638 / 0 | 294 / 0 | 77 / 0 | 1294 | 0 | 0 | 0 |
| $1.25$ | 113 / $-.625$ | 860 / 0 | 307 / .875 | 44 / 2.125 | 1324 | 1655 | 2069 | 292 |
| $9.25$ | 3 / $-4.625$ | 232 / 0 | 70 / 6.475 | 3 / 15.725 | 308 | 2849 | 26353 | 487 |
| **f** | 1031 | 2424 | 1039 | 433 | 4927 | 3604 | 28827 | 568 |
| **fu** | $-516$ | 0 | 727 | 736 | 947 | | | |
| **fu²** | 258 | 0 | 509 | 1251 | 2018 | | | |

Taking the figures from the table,

$$\bar{u} = \frac{947}{4927} = .19$$

$$\bar{v} = \frac{3604}{4927} = .73$$

$$\sigma_x^2 = \frac{2018}{4927} - (.19)^2 = .41 - .04 = .37$$

$$\sigma_x = .61$$

$$\sigma_y^2 = \frac{28827}{4927} - (.73)^2 = 5.85 - .53 = 5.32$$

$$\sigma_y = 2.31$$

$$p = \frac{568}{4927} - (.19)(.73)$$

$$= .12 - .14 = -.02$$

$$r = \frac{-.02}{.61 \times 2.31} = -.014$$

showing practically no correlation.

Example 11.7. The following table shows the distribution of 200 individual items.

| y \ x | −1 | 0 | 1 |
|---|---|---|---|
| 1 | 10 | 20 | 30 |
| 0 | 20 | 40 | 20 |
| −1 | 30 | 20 | 10 |

Calculate the lines of regression and the coefficient of correlation.
*Solution.*

| y \ x | −1 | 0 | 1 | f | fy | fy² | fxy |
|---|---|---|---|---|---|---|---|
| 1 | 10 | 20 | 30 | 60 | 60 | 60 | 20 |
| 0 | 20 | 40 | 20 | 80 | 0 | 0 | 0 |
| −1 | 30 | 20 | 10 | 60 | −60 | 60 | 20 |
| f | 60 | 80 | 60 | 200 | 0 | 120 | 40 |
| fx | −60 | 0 | 60 | 0 | | | |
| fx² | 60 | 0 | 60 | 120 | | | |

The equations for the line of regression of $y$ on $x$ are

$$0 = 120a_0 + 0a_1$$
$$40 = 0a_0 + 120a_1$$

giving

$$a_0 = 0, \text{ and } a_1 = \frac{1}{3}$$

The line of regression of $y$ on $x$ is

$$y = \frac{1}{3}x$$

Similarly, the line of regression of $x$ on $y$ is

$$x = \frac{1}{3}y$$

The coefficient of correlation is

$$\sqrt{\frac{1}{3} \times \frac{1}{3}} = \frac{1}{3}$$

or by Formula 11.1

$$\frac{40}{\sqrt{120 \times 120}} = \frac{1}{3}$$

**11.7. Standard Error of Estimation, Coefficients of Determination and Non-Determination.** The two basic concepts so far discussed in this chapter are the *regression lines* and the *coefficient of correlation.* Associated with these two concepts are

(1) Standard Error of Estimation
(2) Coefficient of Determination
(3) Coefficient of Non-Determination

The formula for the least square regression line of $y$ on $x$ is

$$y = a_0 + a_1 x$$

This is an estimate of the values of $y$ for each $x$. The sum of the squares of the differences between actual and estimated $y$'s is

$$\Sigma (y - y_{\text{est}})^2$$

and the square root of their average value is

$$s_{y \cdot x} = \frac{\Sigma (y - y_{\text{est}})^2}{n} \tag{11.4}$$

This is called the *standard error of estimate* of $y$ on $x$. Similarly, the standard error of estimate of $x$ on $y$ is

$$s_{x \cdot y} = \frac{\Sigma (x - x_{\text{est}})^2}{n}$$

Formula 11.4 may be written for

$$s_{y \cdot x}^2 = \frac{\Sigma y^2 - a_0 \Sigma y - a_1 \Sigma xy}{n} \tag{11.5}$$

The standard error of estimation of $y$ on $x$ is not normally equal to that of $x$ on $y$. These errors have properties similar to the standard deviation and if $n$ is large, bands one, two, and three times the standard error, measured each side of the line of regression, will contain 68%, 95%, and 99.7% of the sample points.

For any value $y$ we may write

$$y - \bar{y} = (y - y_{\text{est}}) + (y_{\text{est}} - \bar{y})$$

where $\bar{y}$ is the mean value of $y$.

It can be shown that

$$\Sigma (y - \bar{y})^2 = \Sigma (y - y_{\text{est}})^2 + \Sigma (y_{\text{est}} - \bar{y})^2$$

The first term on the right is called the *unexplained variation* and the second term, the *explained variation.* The *coefficient of determination* is

$$\frac{\text{explained variation}}{\text{total variation}} = \frac{\Sigma\,(y_{\text{est}} - \bar{y})^2}{\Sigma\,(y - \bar{y})^2}$$

Similarly, the coefficient of *non-determination* is

$$\frac{\Sigma\,(y - y_{\text{est}})^2}{\Sigma\,(y - \bar{y})^2}$$

This coefficient of determination is always positive, and for linear correlation it *is the same whether it is calculated for y on x or for x on y.* It is the square of the coefficient correlation.

$$r = \sqrt{\frac{E}{T}} \quad \text{or} \quad \sqrt{1 - \frac{U}{T}}$$

where

$r$ = coefficient of correlation
$E$ = explained variation
$U$ = unexplained variation
$T$ = total variation

Considering a single value of $y$, understanding of the above will be helped by the illustration in Figure 11.2. It must be re-

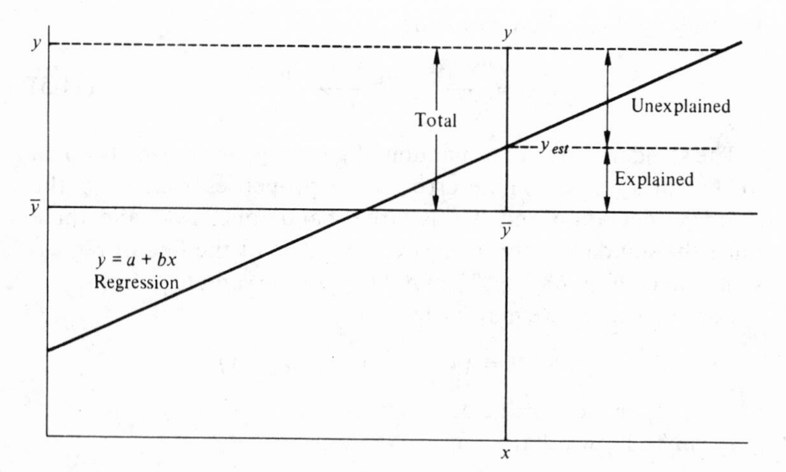


Figure 11.2. Regression. Explained and unexplained variation.

membered that in the equations above we are concerned with the *sum of the squares* of the individual variations.

EXAMPLE 11.8. Using the data in Example 11.7, calculate
(a) The standard error of estimation of $y$ on $x$
(b) The standard error of estimation of $x$ on $y$
(c) The coefficient of determination
(d) The coefficient of non-determination

*Solution.* The line of regression of $y$ on $x$ is

$$y = \frac{1}{3}x$$

Hence, we have

| **x** | **−1** | **0** | **1** | **Total** |
|---|---|---|---|---|
| $\bar{y}$ | 0 | 0 | 0 | |
| $y_{est}$ | $-\frac{1}{3}$ | 0 | $\frac{1}{3}$ | |
| $y_{est} - \bar{y}$ | $-\frac{1}{3}$ | 0 | $\frac{1}{3}$ | |
| $(y_{est} - \bar{y})^2$ | $\frac{1}{9}$ | 0 | $\frac{1}{9}$ | |
| **f** | 60 | 80 | 60 | 200 |
| $f(y_{est} - \bar{y})^2$ | $\frac{60}{9}$ | 0 | $\frac{60}{9}$ | 13.33 |

From the solution to Example 11.7

$$fy^2 = 120$$

and since $\bar{y} = 0$,

$$f(y - \bar{y})^2 = 120$$

$$\text{coefficient of determination} = \frac{13.33}{120} = \frac{1}{9}$$

The coefficient of non-determination can be calculated from a table similar to the above, or as

$$1 - (\text{coefficient of determination}) = 1 - \frac{1}{9} = \frac{8}{9}$$

$$\text{coefficient of correlation} = \sqrt{\frac{1}{9}} = \frac{1}{3}$$

**11.8. Small Volumes of Data.** If we have only two points $(x_1, y_1)$ and $(x_2, y_2)$, a straight line can be drawn through the two points and this would be the line of regression of $x$ on $y$ and of $y$ on $x$. The coefficient of correlation would be 1. This shows that the formula developed above applies only to a reasonable volume of data and cannot be used without adjustment if the number of points ($n$) is small. Just as for the standard deviation when $n$ is small, a factor $\sqrt{\frac{n}{n-1}}$ is introduced, so for correlation we need a factor $\sqrt{\frac{n}{n-2}}$ for small data.

**11.9. Spurious or Nonsense Correlation.** It does not follow that a high degree of correlation means that there is any direct dependence of one variable on the other. Many examples of high degree of correlation can be found for two variables in time series, without any direct connection between the two; such correlation is called spurious or nonsense correlation. While spurious correlation is most common in time series, it can occur in other statistics.

EXAMPLE 11.9. Give examples of spurious correlation.
*Solution.* We can take any two variables which are increasing with time. Thus, we could find positive correlation between the population of Cuba and the number of miles flown by passenger airlines. Again, there would be positive correlation between the cost of living and the number of students entering college. As an example of negative correlation, we could relate the number of sleeping car tickets sold, with the average price of a cinema ticket.

**11.10 Rank Correlation.** In comparing two educational or other tests applied to a small body of students, rank correlation is often useful. The students are ranked in order of their achievement in the two lists, and the difference in rank between the tests $D$ is calculated for each student. $\rho$, the measure of rank correlation is defined by the equation

$$\rho = 1 - \frac{6\Sigma D^2}{n(n^2 - 1)}$$

This is *Spearman's Formula* for rank correlation.

EXAMPLE 11.10. The following are the scores of a group of 10 students in two tests. Calculate the rank correlation.

| Student | Test 1 | Test 2 |
|---|---|---|
| 1 | 87 | 80 |
| 2 | 80 | 95 |
| 3 | 32 | 32 |
| 4 | 65 | 72 |
| 5 | 55 | 67 |
| 6 | 35 | 47 |
| 7 | 82 | 82 |
| 8 | 90 | 85 |
| 9 | 38 | 45 |
| 10 | 72 | 79 |

*Solution.*

| | Test 1 | | Test 2 | | | |
|---|---|---|---|---|---|---|
| **Student** | **Grade** | **Rank** | **Grade** | **Rank** | **D** | **$D^2$** |
| 1 | 87 | 2 | 80 | 4 | 2 | 4 |
| 2 | 80 | 4 | 95 | 1 | 3 | 9 |
| 3 | 32 | 10 | 32 | 10 | 0 | 0 |
| 4 | 65 | 6 | 72 | 6 | 0 | 0 |
| 5 | 55 | 7 | 67 | 7 | 0 | 0 |
| 6 | 35 | 9 | 47 | 8 | 1 | 1 |
| 7 | 82 | 3 | 82 | 3 | 0 | 0 |
| 8 | 90 | 1 | 85 | 2 | 1 | 1 |
| 9 | 38 | 8 | 45 | 9 | 1 | 1 |
| 10 | 72 | 5 | 79 | 5 | 0 | 0 |
| | | | | | | 16 |

$$\rho = 1 - \frac{6 \times 16}{10(100 - 1)}$$

$$= 1 - \frac{96}{990}$$

$$= 1 - .097$$

$$= .903$$

**11.11. Method of Dealing with Ties.** If two students score the same in any test, they can both be given the same rank and the next student a rank two places lower (*bracket method*) or they can be given the *mid-rank value.* Thus,

| Student | Score | Rank | |
|---|---|---|---|
| | | **Bracket Method** | **Mid-Rank Method** |
| 1 | 90 | 1 | 1 |
| 2 | 85 | 2 | 2.5 |
| 3 | 85 | 2 | 2.5 |
| 4 | 80 | 4 | 5 |
| 5 | 80 | 4 | 5 |
| 6 | 80 | 4 | 5 |
| 7 | 75 | 7 | 7 |

**11.12. Correlation of Attributes.** The methods described above are not suitable when we wish to find the correlation of attributes which do not lend themselves to numerical classification. Such attributes could be straight or curly hair, light or dark skin, married or single status, etc. Often we are concerned with a fourfold table (2 × 2 classifications) but not necessarily so.

The *coefficient of contingency* is obtained by comparing the actual distribution in the cells with the distribution to be expected if there were no correlation. This will be discussed further in Chapter 14.

**11.13. Other Coefficients of Correlation.** The *coefficient of reliability* is applied to the correlation of two results from the same test applied to students on two different occasions. When correlation is calculated between characteristics of a parent and an offspring it is called the *coefficient of heredity.* Similarly, correlation between husbands and wives is called the *coefficient of assorted mating.*

**11.14. Multiple Correlation, Non-Linear Correlation.** The principles of correlation can be extended to the correlation between three or more variables and the *coefficient of partial correlation* is the correlation between two variables when the other variables are kept constant.

Just as curves other than a straight line can be used in curve fitting, so formulas may be used for the regression lines which are curves. In this case the correlation is stated to be *non-linear.*

## Problems

**Problem 11.1.** Draw a scatter diagram for the following data and calculate the lines of regression of $y$ on $x$ and $x$ on $y$. Ages are at nearest birthday.

| Case | Age of Husband | Age of Wife |
|---|---|---|
| 1 | 30 | 27 |
| 2 | 30 | 31 |
| 3 | 29 | 25 |
| 4 | 29 | 25 |
| 5 | 28 | 26 |
| 6 | 28 | 25 |
| 7 | 27 | 24 |
| 8 | 27 | 23 |
| 9 | 26 | 24 |
| 10 | 26 | 25 |

**Problem 11.2.** What is the centroid of the data in Problem 11.1?
**Problem 11.3.** Express the lines of regression in Problem 11.1 in terms of the centroid.
**Problem 11.4.** Calculate the coefficient of correlation of the data in Problem 11.1 from (a) the lines of regression and (b) by the product-moment formula (ignore the correction for small volume).
**Problem 11.5.** Is the correlation in Problem 11.4 positive or negative, high or low?
**Problem 11.6.** The following table gives the number of persons employed on farms and the average farm wages per month, with room and board, over the period 1961 to 1966. Calculate the correlation between the two, and state if it is positive or negative, weak or strong, real or spurious.

| Year | Employment (in millions) | Wages |
|---|---|---|
| 1961 | 6.9 | $151 |
| 1962 | 6.7 | 155 |
| 1963 | 6.5 | 159 |
| 1964 | 6.1 | 162 |
| 1965 | 5.6 | 171 |
| 1966 | 5.2 | 185 |

**Problem 11.7.** The following table gives the expectation of life for different ages. Ignoring the adjustment for small volumes of

data, calculate the correlation between the age and the expectation of life.

| Age | Expectation | Age | Expectation |
|---|---|---|---|
| 0 | 70 | 40 | 34 |
| 10 | 62 | 50 | 26 |
| 20 | 53 | 60 | 18 |
| 30 | 43 | 70 | 12 |

Is the correlation spurious?

**Problem 11.8.** Prepare a distribution table for the data in Problem 11.1 and calculate the coefficient of correlation using this table.

**Problem 11.9.** The following data refers to a test applied to 200 students at an interval of 1 year.

| | | First Test | | | | |
|---|---|---|---|---|---|---|
| | | 50–59 | 60–69 | 70–79 | 80–89 | 90–100 |
| Second Test | 90–100 | 0 | 0 | 0 | 0 | 0 |
| | 80–89 | 0 | 10 | 20 | 20 | 0 |
| | 70–79 | 10 | 10 | 40 | 10 | 0 |
| | 60–69 | 20 | 30 | 10 | 0 | 0 |
| | 50–59 | 10 | 10 | 0 | 0 | 0 |

Calculate the line of regression of the second test on the first test.

**Problem 11.10.** What is the centroid of the data in Problem 11.9?

**Problem 11.11.** Show that the line of regression of the second test on the first test in Problem 11.9 goes through the centroid.

**Problem 11.12.** Calculate the coefficient of correlation for the data in Problem 11.9.

**Problem 11.13.** What is the coefficient of correlation when the lines of regression are identical?

**Problem 11.14.** In a correlation study involving a large volume of data, the standard error of estimation of $y$ on $x$ is found to be 1.5. How wide a band must be drawn on either side of the regression line of $y$ on $x$ to include approximately 50% of the points in the scatter diagram.

**Problem 11.15.** In measuring the distance between the regression line and the edges of the band referred to in the previous question, is the measurement made at right angles to the line of regression or parallel to the $y$ axis?

**Problem 11.16.** Calculate the standard error of estimation of $y$ on $x$ for the data in Problem 11.1, using Formula 11.5.

**Problem 11.17.** Check the result of Problem 11.16 by calculating the individual values of $(y - y_{est})$.
**Problem 11.18.** Calculate the coefficient of determination from the data in Problem 11.1.
**Problem 11.19.** Calculate the coefficient of correlation for the data in Problem 11.1 from the result of Problem 11.18.
**Problem 11.20.** Calculate the rank correlation of the following two tests, using the mid-rank method.

| | Marks | |
|---|---|---|
| **Student** | **Test 1** | **Test 2** |
| 1 | 80 | 92 |
| 2 | 67 | 70 |
| 3 | 91 | 78 |
| 4 | 80 | 92 |
| 5 | 73 | 74 |
| 6 | 85 | 83 |
| 7 | 80 | 76 |
| 8 | 78 | 76 |

## Solutions

**Problem 11.1.** Use age 28 and age 25 as origin.

| Case | x − 28 | y − 25 | (x − 28)² | (x − 28)(y − 25) | (y − 25)² |
|---|---|---|---|---|---|
| 1 | 2 | 2 | 4 | 4 | 4 |
| 2 | 2 | 6 | 4 | 12 | 36 |
| 3 | 1 | 0 | 1 | 0 | 0 |
| 4 | 1 | 0 | 1 | 0 | 0 |
| 5 | 0 | 1 | 0 | 0 | 1 |
| 6 | 0 | 0 | 0 | 0 | 0 |
| 7 | −1 | −1 | 1 | 1 | 1 |
| 8 | −1 | −2 | 1 | 2 | 4 |
| 9 | −2 | −1 | 4 | 2 | 1 |
| 10 | −2 | 0 | 4 | 0 | 0 |
| | 0 | 5 | 20 | 21 | 47 |

To calculate the line of regression of $y$ on $x$

$$5 = 10a_0 + 0a_1$$
$$21 = 0a_0 + 20a_1$$

giving

$$y - 25 = .5 + 1.05(x - 28), \quad \text{or} \quad 1.05x - y - 3.9 = 0$$

To calculate the line of regression of $x$ on $y$

$$0 = 10b_0 + 5b_1$$
$$21 = 5b_0 + 47b_1$$

giving

$$x - 28 = -.236 + .472(y - 25), \quad \text{or } x - .472y - 15.964 = 0$$

**Problem 11.2.**

$$\frac{\Sigma (x - 28)}{n} = 0, \qquad \frac{\Sigma (y - 25)}{n} = .5$$

The centroid is $(x - 28) = 0$, $(y - 25) = .5$, or male age 28, female age 25.5.

**Problem 11.3.** The line of regression of $y$ on $x$ is

$$(y - 25) = .5 + 1.05(x - 28)$$
$$y - 25.5 = 1.05(x - 28)$$
$$\bar{y} = 1.05\bar{x}$$

The line of regression of $x$ on $y$ is

$$\bar{x} = .472\bar{y}$$

**Problem 11.4.**

(a) $\sqrt{1.05 \times .472} = \sqrt{.496} = .704$

(b) $$\frac{(10 \times 21) - (0 \times 5)}{\sqrt{(10 \times 20 - 0)(10 \times 47 - 25)}} = \frac{210}{\sqrt{200 \times 445}} = \frac{210}{298} = .705$$

The difference is due to rounding.

**Problem 11.5.** The correlation is positive and fairly high.

**Problem 11.6.** Use origin of 6.0 million and \$160.

| Year | u | v | $u^2$ | uv | $v^2$ |
|---|---|---|---|---|---|
| 1961 | .9 | −9 | .81 | −8.1 | 81 |
| 1962 | .7 | −5 | .49 | −3.5 | 25 |
| 1963 | .5 | −1 | .25 | −.5 | 1 |
| 1964 | .1 | 2 | .01 | +.2 | 4 |
| 1965 | −.4 | 11 | .16 | −4.4 | 121 |
| 1966 | −.8 | 25 | .64 | −20.0 | 625 |
| | 1.0 | 23 | 2.36 | −36.3 | 857 |

Using the product-moment formula

$$r = \frac{-(6 \times 36.3) - (1.0 \times 23)}{\sqrt{[6 \times 2.36 - (1.0)^2][6 \times 857 - (23)^2]}}$$

$$= \frac{-217.8 - 23}{\sqrt{13.16 \times 4613}}$$

$$= \frac{-240.8}{\sqrt{60707}} = \frac{-240.8}{246.5} = -.98$$

The correlation is negative, strong, and spurious.

**Problem 11.7.** Use origins of 30 years of age and 40 years of expectation. Use units of 10 years.

| u | v | u² | uv | v² |
|---|---|---|---|---|
| −3 | 3.0 | 9 | −9.0 | 9.00 |
| −2 | 2.2 | 4 | −4.4 | 4.84 |
| −1 | 1.3 | 1 | −1.3 | 1.69 |
| 0 | 0.3 | 0 | 0 | 0.09 |
| 1 | −0.6 | 1 | −0.6 | 0.36 |
| 2 | −1.4 | 4 | −2.8 | 1.96 |
| 3 | −2.2 | 9 | −6.6 | 4.84 |
| 4 | −2.8 | 16 | −11.2 | 7.84 |
| 4 | −0.2 | 44 | −35.9 | 30.62 |

Coefficient of correlation

$$r = \frac{-(8 \times 35.9) + (4 \times 0.2)}{\sqrt{[(8 \times 44) - 16][(8 \times 30.62) - .04]}}$$

$$= \frac{-287.2 + 0.8}{\sqrt{336 \times 244.9}}$$

$$= \frac{-286.4}{286.9} = -1.00$$

The correlation is very strongly negative. It is not spurious.

**Problem 11.8.** Using origin of male age 28, and female age 25,

| | Husband Age | 26 | 27 | 28 | 29 | 30 | | | | |
|---|---|---|---|---|---|---|---|---|---|---|
| Wife Age | v \ u | −2 | −1 | 0 | 1 | 2 | f | fv | $fv^2$ | fuv |
| 31 | 6 | | | | | 1<br>12 | 1 | 6 | 36 | 12 |
| 30 | 5 | | | | | | | | | |
| 29 | 4 | | | | | | | | | |
| 28 | 3 | | | | | | | | | |
| 27 | 2 | | | | | 1<br>4 | 1 | 2 | 4 | 4 |
| 26 | 1 | | | 1<br>0 | | | 1 | 1 | 1 | 0 |
| 25 | 0 | 1<br>0 | | 1<br>0 | 2<br>0 | | 4 | 0 | 0 | 0 |
| 24 | −1 | 1<br>2 | 1<br>1 | | | | 2 | −2 | 2 | 3 |
| 23 | −2 | | 1<br>2 | | | | 1 | −2 | 4 | 2 |
| | f | 2 | 2 | 2 | 2 | 2 | 10 | 5 | 47 | 21 |
| | fu | −4 | −2 | 0 | 2 | 4 | 0 | | | |
| | $fu^2$ | 8 | 2 | 0 | 2 | 8 | 20 | | | |

Using the product-moment formula

$$r = \frac{(10 \times 21) - (0 \times 5)}{\sqrt{[(10 \times 20) - 0][(10 \times 47) - 25]}} = .705$$

**Problem 11.9.** Use origins of $64\frac{1}{2}$ for each test and intervals of 10 marks, ($u$ = 1st test, $v$ = 2nd test)

| v \ u | −1 | 0 | 1 | 2 | f | fv | $fv^2$ | fuv |
|---|---|---|---|---|---|---|---|---|
| 2 | 0<br>−2 | 10<br>0 | 20<br>2 | 20<br>4 | 50 | 100 | 200 | 120 |
| 1 | 10<br>−1 | 10<br>0 | 40<br>1 | 10<br>2 | 70 | 70 | 70 | 50 |
| 0 | 20<br>0 | 30<br>0 | 10<br>0 | 0<br>0 | 60 | 0 | 0 | 0 |
| −1 | 10<br>1 | 10<br>0 | 0<br>−1 | 0<br>−2 | 20 | −20 | 20 | 10 |
| f | 40 | 60 | 70 | 30 | 200 | 150 | 290 | 180 |
| fu | −40 | 0 | 70 | 60 | 90 | | | |
| $fu^2$ | 40 | 0 | 70 | 120 | 230 | | | |

The equations for the lines of regression of $v$ on $u$ are

$$150 = 200a_0 + 90a_1$$
$$180 = 90a_0 + 230a_1$$

Giving

$$a_0 = .48, \qquad a_1 = .59$$

and

$$(y - 64\tfrac{1}{2}) = .48 + .59(x - 64\tfrac{1}{2})$$

**Problem 11.10.**

$$\sum \frac{u}{n} = \frac{90}{200} = .45$$

$$\sum \frac{v}{n} = \frac{150}{200} = .75$$

The centroid is age

$$64.5 + (10 \times .45) \text{ and } 64.5 + (10 \times .75)$$
$$= 69 \text{ and } 72.$$

**Problem 11.11.** Substituting in the solution to Problem 11.9.

$$.75 = .48 + (.59 \times .45)$$
$$.75 = .48 + .27$$

proving that the line of regression goes through the centroid. It will be recalled that the lines of regression meet at the centroid.

**Problem 11.12.**

$$\frac{(200 \times 180) - (90 \times 150)}{\sqrt{[(200 \times 230) - (90)^2][(200 \times 290) - (150)^2]}}$$

$$= \frac{36000 - 13500}{\sqrt{(46000 - 8100)(58000 - 22500)}}$$

$$= \frac{22500}{\sqrt{(37900)(35500)}}$$

$$= \frac{22500}{36680}$$

$$= .61$$

**Problem 11.13.** The correlation is 1.

**Problem 11.14.** Since the distribution of the points follows the normal distribution, the width of the band on either side of the line of regression must be

$$.67 \times 1.5 = 1.0 \text{ approx.}$$

**Problem 11.15.** Parallel to the $y$ axis. Compare Figure 11.2.

**Problem 11.16.** From Formula 11.5

$$s_{y \cdot x}^2 = \frac{\Sigma(y - 28)^2 - a_0 \Sigma(y - 25) - a_1 \Sigma(x - 28)(y - 25)}{n}$$

$$= \frac{47 - (.5 \times 5) - (1.05 \times 21)}{10}$$

$$= \frac{47 - 2.5 - 22.05}{10}$$

$$= 2.245$$

$$s_{y \cdot x} = 1.5$$

**Problem 11.17.** The line of regression of $y$ on $x$ is

$$y - 25 = .5 + 1.05(x - 28)$$

| Case | x − 28 | .5 + 1.05(x − 28) | y − 25 | $y - y_{est}$ | $(y - y_{est})^2$ |
|---|---|---|---|---|---|
| 1 | 2 | 2.60 | 2 | −.60 | .36 |
| 2 | 2 | 2.60 | 6 | +3.40 | 11.56 |
| 3 | 1 | 1.55 | 0 | −1.55 | 2.40 |
| 4 | 1 | 1.55 | 0 | −1.55 | 2.40 |
| 5 | 0 | .50 | 1 | +.50 | .25 |
| 6 | 0 | .50 | 0 | −.50 | .25 |
| 7 | −1 | −.55 | −1 | −.45 | .20 |
| 8 | −1 | −.55 | −2 | −1.45 | 2.10 |
| 9 | −2 | −1.60 | −1 | +.60 | .36 |
| 10 | −2 | −1.60 | 0 | +1.60 | 2.56 |
| | | | | | 10) 22.44 |
| | | | | | 2.244 |

$$s_{y \cdot x}^2 = 2.244$$

$$s_{y \cdot x} = 1.5$$

**Problem 11.18.**

$$\bar{y} = .5$$

$$\Sigma(y - \bar{y})^2 = 44.5$$

$$\text{Explained variation} = \Sigma(y_{est} - \bar{y})^2 = 44.5 - 22.45 = 22.05$$

$$\text{Coefficient of determination} = \frac{22.05}{44.5} = .496$$

**Problem 11.19.** Coefficient of correlation is $\sqrt{.496} = .704$

**Problem 11.20.**

| | Rank | | | |
|---|---|---|---|---|
| **Student** | **1st Test** | **2nd Test** | **D** | **$D^2$** |
| 1 | 4 | $1\frac{1}{2}$ | $2\frac{1}{2}$ | 6.25 |
| 2 | 8 | 8 | 0 | 0 |
| 3 | 1 | 4 | 3 | 9.00 |
| 4 | 4 | $1\frac{1}{2}$ | $2\frac{1}{2}$ | 6.25 |
| 5 | 7 | 7 | 0 | 0 |
| 6 | 2 | 3 | 1 | 1.00 |
| 7 | 4 | $5\frac{1}{2}$ | $1\frac{1}{2}$ | 2.25 |
| 8 | 6 | $5\frac{1}{2}$ | $\frac{1}{2}$ | 0.25 |
| | | | | 25.00 |

$$\rho = 1 - \frac{6 \times 25}{8 \times 63}$$

$$= 1 - \frac{150}{504}$$

$$= 1 - .30$$

$$= .70$$

# 12

# Sampling Theory

**12.1. Introduction.** The whole of a particular body of data is called the *population* or *universe*, and a representative portion of this body is called a *sample.* Samples play a very important part in statistical work because it is often impossible or too expensive to analyse the whole population. Information obtained from a sample or a set of samples is useful in the estimation of the unknown population *parameters*, such as the mean, the variance, etc. This is called *statistical inference* or *estimation.* (See Chapter 13). Again, we often wish to compare two samples from the same population to determine the *hypothesis* that certain differences are *significant* or not. This is part of *decision theory* (Chapter 14).

EXAMPLE 12.1. Give examples where it is impossible, impractical, or too expensive to study the whole of a population or universe.

*Solution.*

(1) In the decennial census of the United States, an attempt is made to enumerate every person and obtain certain basic information such as age, sex, dependents, etc. However, more detailed information relating to such items as income, housing, etc., is obtained from a sample study.

(2) Between censuses, many sample studies are made on such matters as unemployment, population movements, etc.

(3) In the analysis of the production of items from a factory, sample batches are usually tested for conformity to standards rather than the total output. This saves expense.

(4) In biology and other sciences, it is not normally possible to study every item of the population and sample studies are the only available procedure.

**12.2. Large and Small Sample Theory.** It will be recalled that in calculating a standard deviation and in correlation, certain adjustments had to be made to the formulas when the data was small ($N < 30$). This is equally true in sampling theory. The formulas used for large samples do not apply exactly when N < 30 and the *small sampling theory* has then to be used. The formulas developed in small sampling theory actually apply to samples of all sizes, but they are complex and the more simple formulas for large samples are used whenever practical. The early part of this chapter will be concerned with large samples, and small samples will be discussed in the latter sections.

**12.3. Types of Samples.** There are several different ways of drawing samples from a population.

(1) *Random sampling.* Each member of the population has an equal chance of being chosen.

(2) *Stratified sampling.* A heterogeneous population may be divided into homogeneous subgroups, and the sample is then drawn from each subgroup in a random manner. The proportions of the subgroups in the sample should equal the proportions of the subgroups in the population.

(3) *Judgment sampling.* This is the deliberate selection of a sample by the statistician, to obtain a representative cross section of the population. This method is often used in the construction of a model to represent the population. The techniques of this chapter do not apply to a judgment sample.

A number of other terms are used to represent variants of these three major divisions, such as, systematic, double, sequential, area, cluster, quota, and proportional.

**12.4. Methods of Obtaining Random Samples.** In very many problems, each unit has, or can be assigned, a number. People have Social Security numbers, city houses have street numbers, and automobiles have serial numbers and license numbers. The output from a machine can be numbered serially. If each number in our population were written on a slip of paper, and if all the slips were placed in an urn and mixed thoroughly, then by drawing slips from the urn, a random sample of any desired size could be obtained. Many cases can be thought of where this theoretical idea would not be practical, for example, where the whole population is large or innumerable. Often a sample can be obtained by selecting every number with a last digit 4 (say) or with the last two digits 34( say) in the serial number. It is necessary to determine that selection in this way will not involve *bias*, and when this is suspected, a *table of random numbers* should be used instead.

EXAMPLE 12.2. If it were desired to obtain a sample of all telephones in a city, what would be wrong with selecting all numbers ending in the two selected digits (say 34)?
*Solution.* If a sample of private subscribers were needed, this method would probably prove quite satisfactory, but if a sample of all telephones were desired, this would exclude many large firms whose numbers end in 000.

**12.5. Tables of Random Numbers.** Tables of random numbers are usually tabulated in blocks of 5 digits. A start may be made with any block on any page. If the serial numbers of the units to be sampled ran from 1 to 700, the last three digits of each block of five digits would be used. Any random number greater than 700 would be ignored, and 1 would be treated as 001, 35 as 035, etc.

EXAMPLE 12.3. How would you use a table of random numbers to select a sample from a population where the serial numbers were six digits?
*Solution.* Successive pairs of 5 digit numbers from the table would be used, and the first four digits of each ten digit pair would be ignored.

**12.6. Sampling with or without Replacement.** In the urn procedure described above, each slip drawn from the urn could be replaced after the number on it had been recorded. This gives a random selection process which permits the same number to be selected more than once. Any procedure which does this is called *sampling with replacement*. Provided the population is large, the matter is unimportant, but with a small population, the difference is important. Sampling with replacement enables formulas appropriate for infinite populations to be used.

**12.7. Distribution of Sample Means.** A number of samples, all of size $N$, are taken from a certain population, and the mean of each sample is calculated. We then have a new distribution—the distribution of the means of the samples. These sample means have a *normal distribution*, provided the sample size ($N$) is large, even though the population may not have a normal distribution. The mean of this distribution is $\mu_p$, the mean of the population; and the standard deviation is $\sigma_p/\sqrt{N}$, the standard deviation of the population divided by the square root of the sample size. This standard deviation is called the *standard error* of the sampling distribution of the means.

EXAMPLE 12.4. A population consists of all numbers from 0 to 99. Samples of 5 numbers are selected by means of a table of random numbers, as follows:

51, 77, 27, 46, 40
42, 33, 12, 90, 44
46, 62, 16, 28, 98
93, 58, 20, 41, 86
19, 64, 08, 70, 56

Calculate the means of these samples, and the mean and standard deviation of these sample means.

*Solution.*

Adding the number in each sample and dividing by 5, the means of the five samples are

48.2, 44.2, 50.0, 59.6, 43.4

The mean of the means is

$$\frac{48.2 + 44.2 + 50.0 + 59.6 + 43.4}{5} = 49.1$$

The variance is

$$\frac{(.9)^2 + (4.9)^2 + (.9)^2 + (10.5)^2 + (5.7)^2}{5}$$

$$= \frac{.81 + 24.01 + .81 + 110.25 + 32.49}{5}$$

$$= 33.7$$

$$\text{Standard Deviation} = \sqrt{33.7} = 5.8$$

EXAMPLE 12.5. A population has a mean of 50 and a standard deviation of 30. If a large number of samples each of size 36 are selected, what will be the mean and the standard deviation of the means of the samples.

*Solution.*

$$\text{Mean} = 50$$

$$\text{Standard Deviation} = \frac{30}{\sqrt{36}} = \frac{30}{6} = 5$$

**12.8. Other Sampling Distributions.** Consider a proportion $p$ and a large population, obtained by rolling a die or other means, based on the proportion. If samples are taken from this population, the *sampling distribution of proportion of successes* will be $p$ and the standard deviation (standard error) of the distribution will be

$$\sqrt{\frac{p(1-p)}{N}} = \sqrt{\frac{pq}{N}}$$

where $q = 1 - p$.

Although the population is a binomial distribution, the sampling distribution of the proportion is close to normal.

If two independent sets of samples are taken from two separate populations with means $\mu_1$ and $\mu_2$ and standard deviations $\sigma_1$ and $\sigma_2$ then the mean of the *sums of the means* will be

$$\mu_1 + \mu_2$$

and the mean of the *differences of the means* will be

$$\mu_1 - \mu_2$$

In either case, the standard deviation (standard error) of the distribution of the sums or of the differences of the means will be

$$\frac{\sigma_1^2}{N_1} + \frac{\sigma_2^2}{N_2}$$

where $N_1$ and $N_2$ are the sizes of the samples.

For large $N$ the *sampling distribution* of the standard deviation of samples is nearly normal and its standard error is

$$\frac{\sigma}{\sqrt{2N}}$$

EXAMPLE 12.6. Two sets of samples of size 30 and 50 are taken from the population mentioned in Example 12.5. What are

(1) the means and standard deviations of the means of the two sets of samples
(2) the mean and standard deviation of the sampling distribution of the sums of the means
(3) the mean and standard deviation of the sampling distribution of the difference of the means.

*Solution.*

| | **Set 1** | **Set 2** |
|---|---|---|
| **Mean** | 50 | 50 |
| **Standard Deviation** | $\frac{30}{\sqrt{30}} = 5.5$ | $\frac{30}{\sqrt{50}} = 4.2$ |

*Sum of Sample Means*

$$\text{Mean} = 100$$

$$\text{Standard Deviation} = \sqrt{\frac{30^2}{30} + \frac{30^2}{50}} = \sqrt{30 + 18} = 6.9$$

*Difference of Sample Means*

$$\text{Mean} = 0, \text{Standard Deviation} = 6.9$$

EXAMPLE 12.7. Population $A$ consists of the numbers 3 and 5 distributed in equal proportions. Population $B$ consists of the numbers 1 and 5 distributed in equal proportions. Both populations are infinite. A set of samples $X$ of size 50, is taken from population $A$. They will generally have an approximately equal number of 3's and 5's but any distribution of 3's and 5's totalling 50 is possible. A set of samples, $Y$ of size 100 is taken from population $B$.

A new set of samples is formed by combining the mean of any one of sample $X$ with the mean of any of sample $Y$. What is the mean and standard deviation of this distribution?

*Solution.* For population $A$, the mean is 4 and the standard deviation is 1. For population $B$, the mean is 3 and the standard deviation is 2. The mean of the distribution will be

$$\mu_1 + \mu_2 = 4 + 3 = 7$$

The standard deviation will be

$$\sqrt{\frac{\sigma_1^2}{N_1} + \frac{\sigma_2^2}{N_2}} = \sqrt{\frac{1}{50} + \frac{2}{100}} = \sqrt{\frac{2}{50}} = \frac{1}{5}$$

**12.9. Correction for Finite Population.** If the sample size is $N$ and the population size is $M$, the mean of the sample distribution of means is still equal to the mean of the population,

$$\mu = \mu_p$$

But for the standard deviation

$$\sigma^2 = \frac{\sigma_p^2}{N} \cdot \frac{M - N}{M - 1}$$

EXAMPLE 12.8. What is the correction factor to be applied to the standard deviation for a finite population where the population is 100 and the sample size is 10?
*Solution.* The correction factor to the variance ($\sigma^2$) is

$$\frac{M - N}{M - 1} = \frac{100 - 10}{100 - 1} = \frac{90}{99} = .91$$

The factor to be applied to the standard deviation is

$$\sqrt{.91} = .95$$

EXAMPLE 12.9. If the population is 100, what size of sample corresponds to a correction factor to the standard deviation of .9?
*Solution.*

$$\frac{100 - N}{100 - 1} = (.9)^2$$

Hence

$$100 - N = 99 \times .81$$

$$N = 100 - 80$$

$$N = 20$$

**12.10. Student's *t*-distribution.** It was stated earlier that if the sample size is large, the means of the samples have a normal distribution, even if the population itself is not normal. Even for small samples this is true *if the population has a normal distribution.*

Expressed mathematically,

$$z = \frac{\bar{x} - \mu}{\sigma / N^{1/2}}$$

is a standard normal curve, where $\mu$ and $\sigma$ refer to the population. In most cases $\sigma$ is unknown and we must substitute

$$\sigma_{est} = \sqrt{\frac{N}{N - 1}}\, s$$

where $s$ is the standard deviation of the sample. (See Chapter 4, Section 4.10). The equation

$$t = \frac{\bar{x} - \mu}{s/(N - 1)^{1/2}}$$

is called *Student's t-distribution* and approximates to the normal distribution when $N$ is large.

The $t$-distribution is not unlike the normal distribution, but for the same area under the curve and the same standard deviation, the peak is lower and the tails are higher. The use of tables of the $t$-distribution involve the idea of *degrees of freedom.* Expressed simply the number of degrees of freedom is the size of the sample $N$, minus the number $k$ of *population parameters* (constraints) which must be estimated from the sample observations.

$$\nu = N - k$$

For example, one constraint is involved in estimating the mean since

$$\Sigma (x_i - \bar{x}) = 0$$

**12.11. Probable Error.** The table for the areas under the normal curve (Chapter 7) enable us to determine the probability of values being within any particular range outside the mean. Thus, from the range $-\sigma$ to $+\sigma$ the probability is 68%, from $-2\sigma$ to $+2\sigma$ it is 95.5%, and from $-3\sigma$ to $+3\sigma$ it is 99.7%. The range corresponding to 50% is called the probable error, since values are equally likely to be inside or outside this range. For the normal curve, this range is $-.6745\sigma$ to $+.6745\sigma$.

For the $t$-distribution, this range is larger. For 10 degrees of freedom the range is $-.700\sigma$ to $+.700\sigma$ and for 5 degrees of freedom the range is $-.727\sigma$ to $+.727\sigma$.

**12.12. Table of the *t*-distribution.** In using the *t*-distribution, we are normally concerned with the probability that a given value will be outside the range $-x\sigma$ to $+x\sigma$. The probabilities are tabulated in the following form.

| Degrees of Freedom | Probability 0.50 | 0.10 | 0.05 | 0.01 |
|---|---|---|---|---|
| 1 | 1.000 | 6.31 | 12.71 | 63.66 |
| 2 | 0.816 | 2.92 | 4.30 | 9.92 |
| 3 | .765 | 2.35 | 3.18 | 5.84 |
| 4 | .741 | 2.13 | 2.78 | 4.60 |
| 5 | .727 | 2.02 | 2.57 | 4.03 |
| 10 | .700 | 1.81 | 2.23 | 3.17 |
| 20 | .687 | 1.72 | 2.09 | 2.84 |
| ∞ | .674 | 1.64 | 1.96 | 2.58 |

EXAMPLE 12.10. If the number of degrees of freedom is 10, what range of values will include 90% of the total number of means recorded in a large number of sample tests?
*Solution.* If 90% of the values are within the range, 10% will be outside the range. Entering the table with probability of .10 and 10 degrees of freedom we obtain a value of 1.81. Hence the required range is

$$-1.81\sigma \text{ to } +1.81\sigma$$

## Problems

**Problem 12.1.** A library records in the back of its books the date each book is borrowed. It wishes to determine the average number of times a book is borrowed a year. It is suggested that every tenth book on the shelves be taken down and the number of withdrawals in the last 12 months counted. What is wrong with this sampling technique?
**Problem 12.2.** A public opinion analyst obtains the opinions of passersby at a busy street corner to try to determine an election result. What would be wrong with this procedure?
**Problem 12.3.** At a customs inspection counter at a certain airport, large numbers, 0 to 9, are placed at equal intervals above the counter and baggage is distributed under the numbers according to the next-to-last digit of the baggage ticket. Is this

done to provide an equal distribution of the luggage along the counter? Why is the next-to-last digit used?

**Problem 12.4.** A population consists of five numbers 1, 2, 3, 4, 5. How many different samples of two numbers can be selected? List them.

**Problem 12.5.** Calculate the mean and standard deviation of the population in Problem 12.4 and the mean and standard deviation of the mean of the samples.

**Problem 12.6.** How would the answer to Problem 12.5 be altered if the samples were drawn with replacement?

**Problem 12.7.** The age of the 5,000 male students in a college has a mean of 20.1 and a standard deviation of 2.6 years. If 10 samples of 100 students each are taken, what would be the expected mean of the sample means and the standard deviation of this mean?

**Problem 12.8.** In the previous problem, if instead of 10 samples of 100 students each, there had been 100 samples of 10 students each, how would the results be affected? (Assume large sample theory applies).

**Problem 12.9.** How would the result of Problem 12.8 be affected if only 10 samples had been taken instead of 100?

**Problem 12.10.** In Problem 12.7, what is the expected number of the 10 samples which would have a mean between 20.0 and 21.0?

**Problem 12.11.** In Problem 12.7, what is the expected number of the 10 samples which would have a mean less than 20.5?

**Problem 12.12.** From a table of random numbers, the following three samples, each of 10 individual digits, are extracted:

5, 2, 4, 3, 0, 6, 1, 0, 2, 9
1, 4, 5, 0, 3, 4, 5, 9, 1, 1
7, 0, 9, 5, 5, 9, 6, 4, 6, 0

What are the means of the three samples? What is the mean and standard deviation of these means?

**Problem 12.13.** What is the theoretical mean and standard deviation of the means of three samples of 10 digits taken at random from the digits 0 to 9?

**Problem 12.14.** A true die is rolled 360 times. What is the probability that a six comes up at least 70 times?

**Problem 12.15.** In the last problem what is the probability of at least 40 sixes?

**Problem 12.16.** Two sample studies each of size 25, out of a very large population, produce means of 18.4 and 17.8 respectively. If the standard deviation of the population is 2.0, what is the standard deviation of the sample distribution of these two means?
**Problem 12.17.** What is the standard deviation of the sample distribution of the difference of the two means in Problem 12.16?
**Problem 12.18.** What is the probability that the two samples in Problem 12.16 did in fact come from the same large population?
**Problem 12.19.** What is the probable error of a distribution with a mean of 10.0 and a standard deviation of 2.0?
**Problem 12.20.** In a *t*-distribution with 10 degrees of freedom, what range of multiples of the standard deviation will include $92\frac{1}{2}\%$ of the values? How does this compare with the normal curve?

## Solutions

**Problem 12.1.** The books on the shelves are not a proper sample of the total books in the library because they will include all the books never borrowed and only a small percentage of the books in frequent demand. A random sample taken from the books on the shelves will not be a random sample of all the books in the library.
**Problem 12.2.** (1) The passersby at a busy street corner are not a random sample of the voting public and (2) such a sample is likely to be too small to be of value.
**Problem 12.3.** The method is likely to produce a reasonable distribution of the luggage and will bring together the luggage of individual passengers since they will usually have the same next-to-last digit.
**Problem 12.4.** From Chapter 6, the number of samples is

$$\binom{5}{2} = \frac{5!}{2! \times 3!} = \frac{5 \times 4}{2} = 10$$

They are

$$(1, 2);\ (1, 3);\ (1, 4);\ (1, 5);\ (2, 3);$$
$$(2, 4);\ (2, 5);\ (3, 4);\ (3, 5);\ (4, 5).$$

**Problem 12.5.**

*Population* Mean = 3 Standard Deviation = $\sqrt{2}$ = 1.4

| Sample | Mean | $x - \bar{x}$ | $(x - \bar{x})^2$ |
|---|---|---|---|
| 1, 2 | 1.5 | −1.5 | 2.25 |
| 1, 3 | 2.0 | −1.0 | 1.00 |
| 1, 4 | 2.5 | −0.5 | .25 |
| 1, 5 | 3.0 | 0 | 0 |
| 2, 3 | 2.5 | −0.5 | .25 |
| 2, 4 | 3.0 | 0 | 0 |
| 2, 5 | 3.5 | 0.5 | .25 |
| 3, 4 | 3.5 | 0.5 | .25 |
| 3, 5 | 4.0 | 1.0 | 1.00 |
| 4, 5 | 4.5 | 1.5 | 2.25 |
| | 10)30.0 | 0 | 10)7.50 |
| | Mean = 3.0 | | Variance = .75 |

Mean of samples = 3

Standard deviation = $\sqrt{.75}$ = .87

**Problem 12.6.** With replacement, the samples would be

(1, 1); (1, 2); (1, 3); etc.

and we must distinguish between (1, 2) and (2, 1) as otherwise we shall give double weight to (1, 1); (2, 2); etc. The mean of the samples will still be 3 but the standard deviation will be calculated as follows:

| Sample | Frequency | Mean | $f(x - \bar{x})$ | $f(x - \bar{x})^2$ |
|---|---|---|---|---|
| 1, 1 | 1 | 1 | −2 | 4 |
| 2, 2 | 1 | 2 | −1 | 1 |
| 3, 3 | 1 | 3 | 0 | 0 |
| 4, 4 | 1 | 4 | 1 | 1 |
| 5, 5 | 1 | 5 | 2 | 4 |
| All Other | 20 | 3 | 0 | 15* |
| | 25 | | 0 | 25)25 |
| | | | | 1 |

Mean of samples = 3

Standard Deviation = $\sqrt{1}$ = 1

*From previous table.

**Problem 12.7.** The mean would be the same as the mean of the population, 20.1, and the standard deviation is

$$\frac{2.6}{\sqrt{100}} = .26$$

**Problem 12.8.** The mean would be unaltered at 20.1. The standard deviation, by the formula for large samples, would be

$$\frac{2.6}{\sqrt{10}} = .82$$

**Problem 12.9.** The results would not be altered in any way. The mean and standard deviation of the sample means is the same if there is one or any number of samples.

**Problem 12.10.** The mean of the means is 20.1 and the standard deviation is .26.

$$20.0 = m - .4\sigma$$

$$21.0 = m + 3.5\sigma$$

From Table 7.1, the area under the normal curve

$$\begin{aligned} \text{between } -.4\sigma \text{ and } 0 &= .155 \\ \text{between } 0 \text{ and } 3.5\sigma &= \underline{.500} \\ & \phantom{=} .655 \ \textit{total} \end{aligned}$$

The probability of a value between 20.0 and 21.0 is .655 and the expected number out of ten samples is 6.55.

**Problem 12.11.** The area under the normal curve between $-\infty$ and $+1.54\sigma$ is .938. The expected number is

$$10 \times .938 = 9.4 \text{ approx.}$$

**Problem 12.12.** The means are 3.2, 3.3, and 5.1. The mean of the means is 3.87. The variance is

$$\frac{(.67)^2 + (.57)^2 + (1.23)^2}{3} = .76$$

The standard deviation is

$$\sqrt{.76} = .87$$

**Problem 12.13.** The mean is 4.5 and the variance is

$$\frac{\frac{1}{10}\sum_{r=o}^{9} r^2 - (4.5)^2}{10} = \frac{\frac{1}{10} \times 285 - 20.25}{10}$$

$$= \frac{28.5 - 20.25}{10} = .825$$

$$\text{The standard deviation} = \sqrt{.825} = .91$$

Since this is an example of sampling with replacement, no adjustment is needed for the small size of the sample.

**Problem 12.14.** This is a sample study of an infinite population of all possible rolls of the dice. The probability of a six is 1/6, so that $p = 1/6$ and $q = 5/6$.

For the sample distribution of the proportion of successes, the mean will be $p$ or 1/6 and the standard deviation is

$$\sqrt{\frac{\frac{1}{6} \times \frac{5}{6}}{360}}$$

The expected number of sixes will be $360 \times 1/6 = 60$ and the standard deviation is

$$\sqrt{360 \times \frac{1}{6} \times \frac{5}{6}} = \sqrt{50} = 7.07$$

Since this is a discrete distribution, we want the probability of at least 69.5. Now $69.5 - 60 = 9.5$ or 1.34 times the standard deviation. Hence, we want the area of the normal curve from $1.34\sigma$ to $\infty$. From Table 7.1 this is

$$.50 - .41, \text{ or } .09 \text{ approximately}$$

**Problem 12.15.** 39.5 is 20.5 from the mean or $2.9\sigma$ from the mean. Therefore we need the area under the normal curve between $-2.9\sigma$ and $+\infty$, or

$$.4981 + .5000 = .998$$

**Problem 12.16.** In each case, the standard deviation is

$$\frac{2.0}{\sqrt{25}} = 0.4$$

**Problem 12.17.**

$$\sqrt{\frac{(2.0)^2}{25} + \frac{(2.0)^2}{25}} = \sqrt{\frac{8}{25}} = 0.57$$

**Problem 12.18.** The difference between the means is 0.6 and the standard deviation is 0.57. Hence, the difference is $1.05\sigma$ and from Table 7.1, the probability is $2 \times .35$ or .7.

**Problem 12.19.**

$$.6745 \times 2, \quad \text{or} \quad 1.35$$

**Problem 12.20.** From the table in Section 12.12, 90% of the values are in the range $\pm 1.81\sigma$ and 95% of the values are in the range $\pm 2.23\sigma$. Interpolating between these values, $92\frac{1}{2}\%$ of the values will lie in the range $\pm 2.0\sigma$. For the normal curve, the range is $\pm 1.78\sigma$.

# 13

# Estimation

**13.1. Introduction.** Often the only information we can obtain about a population is from a study of samples taken at random from it. *Statistical inference*, or statistical *estimation theory*, is the procedure by which population parameters are estimated from a study of samples.

**13.2. Biased and Unbiased Estimates.** It was shown in the last chapter that the theoretical mean of the sampling distribution of the means is equal to the mean of the population. When this is true for any parameter, the statistic is referred to as an *unbiased estimator*.

The theoretical mean of the sampling distribution of variances is equal to

$$\frac{N - 1}{N} \sigma^2$$

where $\sigma^2$ is the population variance, and $N$ is the sample size. Here, a sample variance is a *biased estimator* of the population variance, but an unbiased estimator of

$$\frac{N - 1}{N} \sigma^2$$

Put another way,

$$\frac{N}{N - 1} \sigma_s^2$$

(where $\sigma_s^2$ is a sample variance, or the mean of a number of sample variances) is an unbiased estimator of the population variance $\sigma^2$.

**13.3. The *k*-statistics.** It will be recalled from Chapter 5 that the moments of a distribution about the origin were $m'_1$, $m'_2$, $m'_3$ and $m'_4$ and were 0, $m_2$, $m_3$ and $m_4$ about the mean.

The corresponding moments of an *individual sample* will be approximations to the population moments, but if the sample is not large, better approximations will be the following functions called *k-statistics*.

| Sample k-statistic | Estimate of Population Statistic |
| --- | --- |
| $k_1 = \bar{x} = m'_1$ | $\mu'_1$ |
| $k_2 = N\sigma^2/(N-1) = Nm_2/(N-1)$ | $\mu_2$ |
| $k_3 = N^2 m_3/(N-1)(N-2)$ | $\mu_3$ |
| $k_4$ | $\mu_4 - 3\mu_2^2$ |

where

$$k_4 = \frac{N^2}{(N-1)(N-2)(N-3)}[(N+1)m_4 - 3(N-1)m_2^2]$$

These are unbiased estimators.

Where $N$ is large,

$$k_2 = m_2, \quad k_3 = m_3 \quad \text{and} \quad k_4 = m_4 - 3m_2^2$$

EXAMPLE 13.1. Which of the following are unbiased estimators and which are biased estimators of the population parameters?

1. The mean of a sample as an estimator of the mean of the population.

2. The median of a sample as an estimator of the mean of the population.

3. The variance of a sample as an estimator of the variance of the population.

*Solution.* The mean and median are both unbiased estimators; this was shown for the mean in the previous chapter, and since the sample distribution of the means is a normal curve, it will be equally true of the median. The variance is a biased estimator.

EXAMPLE 13.2. What is an unbiased estimator of the standard deviation of a population?

*Solution.* Since the variance is an unbiased estimator of

$$\frac{N-1}{N}\sigma^2$$

then it follows that the variances times $\frac{N}{N-1}$ is an unbiased estimator of $\sigma^2$ and hence,

$$\sqrt{\frac{N}{N-1}}\,\sigma_s$$

is an unbiased estimator of the standard deviation of the population.

**13.4. Consistent Estimates.** Consistent estimates of a parameter are estimates which become more accurate as the sample size increases. In technical language, an estimate is consistent if the probability that it differs from the true value by less than a given amount, however small, tends to unity as $N \to \infty$. The estimators discussed in this chapter are all consistent.

**13.5. Point and Interval Estimates.** Estimates can be stated in the form of a single value such as 4.32 or a range of values 4.32 ± .13. The first is called a *point estimate*, the latter, an *interval estimate.* Clearly, the interval estimate is more satisfactory because it not only indicates the *probable value* but also the *reliability* of the estimate.

**13.6. Confidence Limits and Intervals.** For large samples ($N \geq 30$), the sampling distribution of the mean is approximately normal, so that we can expect to find any actual mean within the range $\mu - \sigma$ and $\mu + \sigma$ in about 68% of the samples. Using Table 7.1, we can choose any appropriate percentage and determine the range in which we may expect to find the mean of a sample. The chosen percentage is called the *confidence level.* Thus, for 50% confidence level, the *confidence interval* is

$$\mu - .67\sigma \text{ to } \mu + .67\sigma$$

These two values are called the *confidence limits*. The factor .67 is called the *confidence coefficient* ($z_c$). The following table gives the confidence coefficients for given confidence levels in a normal distribution.

**Normal Curve**

| Confidence Level | Confidence Coefficient ($z_c$) | Confidence Level | Confidence Coefficient ($z_c$) |
|---|---|---|---|
| 99.73% | 3.00 | 95% | 1.96 |
| 99% | 2.58 | 90% | 1.64 |
| 98% | 2.33 | 80% | 1.28 |
| 97% | 2.17 | 75% | 1.15 |
| 96% | 2.05 | 68.27% | 1.00 |
| 95.45% | 2.00 | 50% | .67 |

What estimate can we make of the mean of the whole population and what confidence can we attach to it? We have already seen that the mean of the sample means is an unbiased estimator of the population, and for an infinite population, the population mean is

$$\bar{X} \pm z_c \frac{\sigma}{\sqrt{N}} \tag{13.1}$$

where $\bar{X}$ is the sample mean and $\sigma$ is the standard deviation of the population. However, $\sigma$ is unknown and we use instead the sample estimate $\sigma_s$. $N$ is the sample size.

For a finite population, the formula becomes

$$\bar{X} \pm z_c \frac{\sigma}{\sqrt{N}} \sqrt{\frac{N_p - N}{N_p - 1}}$$

where $N_p$ is the population size.

For a small sample ($N < 30$) these formulas will over-estimate the reliability of the estimate and Formula 13.1 becomes,

$$\bar{X} \pm t_c \frac{\sigma}{\sqrt{N - 1}}$$

where $t$ is obtained from Student's $t$-distribution.

EXAMPLE 13.3. (1) Which is the more reliable estimate, 4.32 ± .03 or 4.32 ± .13?

(2) Which is the more reliable estimate, 4.32 ± .13 or 8.32 ± .13?

*Solution.* (1) For a given confidence level, 4.32 ± .03 is more reliable than 4.32 ± .13. If the confidence level is 90%, the former implies that there is a 90% probability that the true value does not depart from the estimate of 4.32 by more than .03, while the latter implies that there is a 90% probability that the true value does not depart from the estimate by more than .13.

(2) Although, in the second example, the confidence intervals have the same breadth in both cases, this breadth, when expressed as a percentage of the estimate, is greater in the former case and hence $4.32 \pm .13$ is less reliable than $8.32 \pm .13$. This will be made clearer when we consider estimates such as

$$1 \pm 1 \quad \text{and} \quad 100 \pm 1$$

The former has the confidence interval of 0 to 2, which does not provide a particularly accurate estimate; the latter has an interval 99 to 101, which is a fairly close estimate.

EXAMPLE 13.4. For a large sample, if the mean is

$$9.8 \pm .07$$

for a 95% confidence level, what is it for an 80% confidence level?
*Solution.* For a 95% confidence level, the confidence coefficient is 1.96. For an 80% confidence level, it is 1.28. Hence, the value for an 80% confidence level is

$$9.8 \pm \frac{1.28}{1.96} \times .07$$

or

$$9.8 \pm .05$$

EXAMPLE 13.5. What is the confidence coefficient for confidence level 97.5%?
*Solution.* From the table, the coefficient for 97% is 2.17, and for 98% it is 2.33. Interpolating, we have for 97.5%

$$\frac{2.17 + 2.33}{2} = 2.25$$

Alternatively, entering Table 7.1 with

$$\frac{.975}{2} = .4875$$

we get a figure of 2.24. The latter is the correct figure, but the former is a reasonable approximation.

EXAMPLE 13.6. If the value of the mean of a population, estimated from a sample of size 100 is

$$8.12 \pm 2.5$$

what would you expect the estimate to be, based on a sample of size 200?

*Solution.* From Formula 13.1 the first estimated value is

$$\bar{X} \pm z_c \frac{\sigma}{\sqrt{100}}$$

and the second estimated value is

$$\bar{X} \pm z_c \frac{\sigma}{\sqrt{200}}$$

We will expect $\bar{X}$ and $\sigma$ to be the same in both cases, hence the new value will be

$$8.12 \pm \frac{\sqrt{100}}{\sqrt{200}} \times 2.5$$

or

$$8.12 \pm 1.8$$

EXAMPLE 13.7. If the probable error of a certain statistic which is distributed normally is 2.0, what is the 90% confidence interval?

*Solution.* It was explained in the previous chapter (Section 12.11) that the probable error is the 50% confidence level which has a confidence coefficient of .67. A 90% confidence level has a confidence coefficient of 1.64. Hence, the 90% confidence interval is the mean value plus or minus

$$2.0 \times \frac{1.64}{.67} = 4.9$$

**13.7. Confidence Intervals for Other Statistics.** We have seen that for a sample size $N$, the confidence intervals for the population mean is

$$\bar{X} \pm z_c \frac{\sigma}{\sqrt{N}}$$

For some common statistics, the confidence intervals are

*Proportions* $$P \pm z_c \sqrt{\frac{pq}{N}} \qquad (13.2)$$

where $P$ is the proportion of successes in a sample size $N$, and $p$ is the population proportion of successes, and $q = 1 - p$. For a large sample, it is generally satisfactory to write $P$ for $p$ giving

$$P \pm z_c \sqrt{\frac{P(1 - P)}{N}}$$

For a finite population, Formula 13.2 becomes

$$P \pm z_c \sqrt{\frac{pg}{N}} \sqrt{\frac{N_p - N}{N_p - 1}}$$

*Difference of Means*

$$\overline{X}_1 - \overline{X}_2 \pm z_c \sqrt{\frac{\sigma_1^2}{N_1} + \frac{\sigma_2^2}{N_2}} \qquad (13.3)$$

*Sums of Means*

$$\overline{X}_1 + X_2 \pm z_c \sqrt{\frac{\sigma_1^2}{N_1} + \frac{\sigma_2^2}{N_2}}$$

*Standard Deviation*

$$s \pm \frac{z_c \sigma_s}{\sqrt{2N}}$$

if the population is normally distributed.

EXAMPLE 13.8. A sample of 200 employees of a large company indicated that 65% thought the company's promotion procedures were satisfactory. Find the 50%, 95% and 99% confidence limits for the proportion of all employees who were satisfied with the promotion procedures.

*Solution.*

$$\sqrt{\frac{pq}{N}} = \sqrt{\frac{.65 \times .35}{200}} = \sqrt{.00114} = .034$$

For 50% confidence limits we have

$$.65 \pm (.67 \times .034)$$

or

$$.65 \pm .02$$

For 95% confidence limits we have

$$.65 \pm (1.96 \times .034)$$

$$.65 \pm .07$$

For 99% confidence limits we have

$$.65 \pm (2.58 \times .034)$$

$$.65 \pm .09$$

EXAMPLE 13.9. A sample of 50 male students in University *A* gives a mean height of 72.1 inches and standard deviation of 3.0 inches. A sample of 100 male students in University *B* gives a mean height of 71.0 inches and a standard deviation of 2.0 inches. Find the 95% confidence limits of the difference in the mean height of the male students of the two universities.

*Solution.* From Formula 13.3 the difference in the means is

$$72.1 - 71.0 = 1.1 \text{ inch}$$

$$z_c\sqrt{\frac{\sigma_1^2}{N_1} + \frac{\sigma_2^2}{N_2}} = 1.96\sqrt{\frac{(3.0)^2}{50} + \frac{(2.0)^2}{100}}$$

$$= 1.96 \times .47$$

$$= 0.9$$

Difference in the means $= 1.1 \pm 0.9$.

**13.8. Choice of Size of Sample.** It can be readily seen that the larger the sample, the narrower is the confidence interval for any specified confidence limits. Hence, in planning a sample study, the size of the sample will depend on the confidence interval and the confidence limits desired. It will also depend on cost and feasibility.

EXAMPLE 13.10. The mean and standard deviation of a large population are approximately 10.0 and 1.0, respectively. What sized sample should be used to determine the mean within a confidence interval $\pm$ 0.1 with 90% confidence?

*Solution.* The confidence interval is

$$\bar{X} + z_c\frac{\sigma}{\sqrt{N}}$$

Hence,

$$0.1 = z_c\frac{1.0}{\sqrt{N}}$$

For 90% confidence,

$$z_c = 1.64$$

$$0.1 = 1.64 \times \frac{1.0}{\sqrt{N}}$$

$$\sqrt{N} = \frac{1.64 \times 1.0}{0.1} = 16.4$$

$$N = (16.4)^2 \text{ or } 269$$

## Problems

**Problem 13.1.** Six scores, chosen at random from a large body of students taking a certain test, are

73, 84, 70, 83, 65, 75

Determine unbiased and efficient estimates of (1) the true mean and (2) the true standard deviation.

**Problem 13.2.** Compare the sample standard deviation with the estimated standard deviation of the population in the previous problem.

**Problem 13.3.** How large must a sample be for the unbiased estimate of the standard deviation of the population to differ from the sample standard deviation by less than $2\frac{1}{2}\%$?

**Problem 13.4.** What is the median of the sample scores in Problem 13.1? Is this an unbiased estimate of the mean of the population? Is this an efficient estimate?

**Problem 13.5.** A sample of size 10 of a certain population is

6.8, 7.0, 7.2, 7.4, 7.5, 7.5, 7.6, 7.8, 8.0, 8.2

Calculate the $k$-statistics of this sample and use them to estimate the parameters of the population.

**Problem 13.6.** What is an estimate of the mean and standard deviation of the means of 20 samples, each of size 10, taken from the population referred to in the previous problem?

**Problem 13.7.** A person, wishing to estimate the true value of a certain stock, takes the price on the New York stock market on January 1st as his first estimate. As his second estimate he takes the mean of the price on January 1st and January 2nd, and as his third estimate he takes the mean of the price on the first three days of the year and so on. Would this provide a consistent estimate?

**Problem 13.8.** A statistic is stated to have a value of 13.8 ± 1.5. What additional information is needed to interpret this statement?

**Problem 13.9.** A true die is rolled. What are the confidence levels for the following statements of the value expected to be observed?

(1) 3.5 ± 1.0
(2) 3.5 ± 2.0
(3) 3.5 ± 3.0

**Problem 13.10.** What is the confidence coefficient for the normal curve corresponding to a 25% confidence level?

**Problem 13.11.** There are 100 students in a particular age group in a school. A random sample of 50 students shows a mean I.Q. of 105 and a standard deviation of 10. At a 50% confidence level, what is the mean I.Q. of the group?

**Problem 13.12.** How would the solution to the above problem be modified if the sample were only 5 students?

**Problem 13.13.** If a certain statistic has a value of 13.2 ± 1.0 at 75% confidence level, what is the confidence level corresponding to a value of 13.2 ± 2.0?

**Problem 13.14.** What is the probable error of the statistic 13.2 referred to in the previous problem?

**Problem 13.15.** A study shows that among 100 persons reaching age 65, 3 die within a year. If these persons can be considered a random sample of the population, what is the confidence limits of the percentage, 3%, as applied to all persons reaching age 65, (1) at 99% confidence level, and (2) at 50% confidence level?

**Problem 13.16.** A poll reports that 54% of the electorate will vote for candidate *A*. If the poll is based on a random sample, and none of the electorate change their mind between the poll and the election, and everyone votes, what is the probability that candidate *A* will obtain at least 50% of the votes if the poll is based on (a) 100 people, (b) 1000 people?

**Problem 13.17.** In Problem 13.16, how large a poll is needed to give a 90% probability of *A* obtaining at least 50% of the votes?

**Problem 13.18.** Random sample statistics of 100 married men and of 200 married women show that they average 12 and 15 days of sickness a year respectively. The standard deviation of these figures are 4 and 6 respectively. Find for 90% confidence limits, the mean number of combined days of sickness of a husband and wife.

**Problem 13.19.** From a random sample study it is found that 10% of men and 70% of women like soap operas. If the studies are based on 100 men and 100 women, what, at the 99.73% confidence level, is the difference between the two percentages?

**Problem 13.20.** What are the confidence levels that the standard deviations in Problem 13.18 do not differ from the true population standard deviations by more than 10%.

## Solutions

**Problem 13.1.** An unbiased and efficient estimate of the mean of the population is the mean of the sample,

$$\frac{73 + 84 + 70 + 83 + 65 + 75}{6} = \frac{450}{6} = 75$$

An unbiased and efficient estimate of the variance of the population is

$$\frac{N}{N-1}\sigma_s^2 = \frac{\Sigma(x - \bar{x})^2}{N-1}$$

$$= \frac{(-2)^2 + (9)^2 + (-5)^2 + (8)^2 + (-10)^2 + (0)^2}{5}$$

$$= \frac{4 + 81 + 25 + 64 + 100}{5} = 55$$

An unbiased and efficient estimate of the standard deviation is

$$\sqrt{55} = 7.4$$

**Problem 13.2.**

$$\text{Sample variance} = \frac{4 + 81 + 25 + 64 + 100}{6} = 46$$

Sample standard deviation $= \sqrt{46} = 6.8$

Estimated standard deviation of population $= 7.4$

**Problem 13.3.** We require to find the value of $N$ such that

$$\sqrt{\frac{N}{N-1}} < 1.025$$

Evaluating for $\sqrt{\frac{N}{N-1}} = 1.025$ we have

$$N = (N - 1)\,1.051$$
$$.051N = 1.051$$
$$N = 20.6$$

For $N = 20.6$, the unbiased estimate of the standard deviation of the population is 2½% greater than the sample standard deviation. If the difference is to be less than 2½%, the sample must be of size 21 or more.

**Problem 13.4.** The two middle values are 73 and 75. The median is therefore 74. This is an unbiased but not an efficient estimate of the mean of the population.

**Problem 13.5.** The sum of the ten values is 75.0 and hence,

$$k_1 = \bar{x} = 7.5$$

Measuring from the mean, we have

| $x - \bar{x}$ | $(x - \bar{x})^2$ | $(x - \bar{x})^3$ | $(x - \bar{x})^4$ |
|---|---|---|---|
| −0.7 | .49 | −.343 | .240 |
| −0.5 | .25 | −.125 | .063 |
| −0.3 | .09 | −.027 | .008 |
| −0.1 | .01 | −.001 | .000 |
| 0 | 0 | 0 | 0 |
| 0 | 0 | 0 | 0 |
| 0.1 | .01 | .001 | .000 |
| 0.3 | .09 | .027 | .008 |
| 0.5 | .25 | .125 | .063 |
| 0.7 | .49 | .343 | .240 |
| 0 | 1.68 | 0 | .622 |

$$m_2 = .168$$
$$m_3 = 0$$
$$m_4 = .062$$

$$k_2 = \frac{10}{9} \times .168 = .19$$

$$k_3 = 0$$

$$k_4 = \frac{100}{9 \times 8 \times 7}[(11 \times .062) - 3 \times 9 \times (.168)^2]$$

$$k_4 = 2.0\,[.68 - .76] = -.16$$

These are unbiased estimators of $\mu_1'$, $\mu_2$, $\mu_3$, and $\mu_4 - 3\mu_2^2$

**Problem 13.6.** The mean of the one sample is 7.5 and this is the only estimate available of the mean of the population. Hence, the estimate of the mean of the means of the 20 samples is 7.5.

The estimate of the standard deviation of the population is $\sqrt{k_2} = \sqrt{.19} = .44$. The standard deviation of the 20 sample means will be

$$\frac{\sigma}{\sqrt{10}} = \frac{.44}{3.16} = .14$$

Remember that $N$ is the size of the sample, not the number of samples. The latter does not affect the result.

**Problem 13.7.** The price of a stock on various days will not provide a homogeneous group of data and the adding of additional data from a time series such as this will not meet the definition of a *consistent* estimate, since there is neither a *true value* to which the series of estimates can converge nor will the estimates approach any selected value in the manner described in Section 13.4.

**Problem 13.8.** We require to know the confidence level used in determining the confidence interval 1.5. Is it a 50% figure, a 90% figure, or what? Sometimes a 50% level is assumed when no limit is stated.

**Problem 13.9.** (1) The value $3.5 \pm 1.0$ includes 3 and 4 only. The probability of the die coming up 3 or 4 is 1/3. Hence, the confidence level is 33⅓%.

For (2) the confidence level is 66⅔% and for (3) it is 100%.

**Problem 13.10.** The value of the confidence coefficient for a confidence level of 25% cannot be obtained from the table in Section 13.6, and reference must be made to Table 7.1 giving the areas under the normal curve. For 25% confidence level, 75% of the values will be outside the limits: 37½% below the confidence limits and 37½% of the values above them. Since the areas tabulated are for one side of the curve only, we need $z$ corresponding to the area $(.50 - .375)$ or .125. This area corresponds to $z = 0.32$. It will be noted that this value of $z$ is approximately 1/2 the value for a 50% confidence level.

**Problem 13.11.** The estimated mean is

$$105 \pm .67 \frac{10}{\sqrt{50}} \sqrt{\frac{100 - 50}{100 - 1}}$$

$$= 105 \pm .67 \frac{10}{\sqrt{50}} \frac{\sqrt{50}}{\sqrt{99}}$$

$$= 105 \pm 0.7$$

**Problem 13.12.** With only 5 students, the sample would be "small" and the formula used in the solution to the previous problem could not be used. Student's $t$-formula would apply.

**Problem 13.13.** We are given that

$$z_c \frac{\sigma}{\sqrt{N}} = 1.0$$

for a 75% confidence level. Hence,

$$\frac{\sigma}{\sqrt{N}} = \frac{1.0}{1.15}$$

For a confidence interval of $\pm 2.0$

$$z_c \frac{1.0}{1.15} = 2.0$$

or

$$z_c = 2 \times 1.15 = 2.3$$

From the table in Section 13.6 this figure corresponds to a confidence level of 98%.

**Problem 13.14.** For the probable error $z_c = .67$, hence, the confidence interval is

$$13.2 \pm \left(\frac{1.0}{1.15} \times .67\right)$$

$$= 13.2 \pm .58$$

The probable error is .58.

**Problem 13.15.** For 99% confidence level, the confidence limits are

$$.03 \pm 2.58 \sqrt{\frac{pq}{N}}$$

$$.03 \pm 2.58 \sqrt{\frac{.03 \times .97}{100}}$$

or

$$= .03 \pm .04$$

For 50% the limits are

$$.03 \pm .67 \sqrt{\frac{.03 \times .97}{100}}$$

$$= .03 \pm .01$$

**Problem 13.16.** The confidence limits are

$$.54 \pm z_c\sqrt{(.54)(.46)/N}$$

For $N = 100$ this gives

$$.54 \pm z_c(.050)$$

The candidate will fail to be elected if his percentage of votes is less than .54 – .04, not if it is greater than .54 + .04. Hence, while we equate

$$.50 = .54 - z_c(.050)$$

to obtain the value of $z_c$, we must not take the resultant confidence level as the probability of election.

$z_c = .8$ in this case, which corresponds to a 58% confidence level. The probability of not being elected is only 1/2 (1 – .58) or .21 and the probability of election is .79.

For $N = 1000$ we have

$$.54 \pm z_c \times .016$$

giving $z_c = 2.5$ when equated to .50, which corresponds to the 99% confidence level. The probability of election is 99½%.

**Problem 13.17.** For a 90% chance of being elected, we need only an 80% confidence level since errors of estimation on the high side will not affect the results.

$$.50 = .54 - 1.28 \times \frac{.50}{\sqrt{N}}$$

$$\sqrt{N} = \frac{1.28 \times .50}{.04} = 16$$

$$N = 256$$

**Problem 13.18.** The confidence limits are

$$27 \pm z_c \sqrt{\frac{\sigma_1^2}{N_1} + \frac{\sigma_2^2}{N_2}}$$

$$= 27 \pm 1.64 \sqrt{\frac{16}{100} + \frac{36}{200}}$$

$$= 27 \pm 1.64 \times .58$$

$$= 27 \pm .95$$

**Problem 13.19.**

$$(.70 - .10) \pm 3.00 \sqrt{\frac{.10 \times .90}{100} + \frac{.30 \times .70}{100}}$$

$$= .60 \pm 3.00 \times .055$$

$$= .60 \pm .16$$

or $\quad 60\% \pm 16\%$

**Problem 13.20.** The percentage error is

$$\frac{z_c \dfrac{\sigma}{\sqrt{2N}}}{\sigma}$$

Equating this to 10%

$$\frac{z_c}{\sqrt{2N}} = .1$$

or, $z_c = .1 \times \sqrt{200}$ for men; $.1 \times \sqrt{400}$ for women. That is, $z_c = 1.41$ for men; 2.00 for women, giving confidence levels of 84% and 95½% respectively.

# 14

# Decision Theory

**14.1. Tests of Hypotheses.** In the application of statistics to the solution of many problems, we make a *statistical hypothesis* or *decision* about the population, and then proceed to *test the decision* from a sample study. If the sample study shows that the observed results differ from what would be expected on our hypotheses to a considerably greater extent than would be expected by mere chance, then we say the difference is *significant*. Procedures which enable such decisions to be made are called *tests of hypotheses* or *tests of significance*.

**14.2. The Null Hypothesis.** One of the most useful procedures in decision theory is to make an hypothesis which we can later reject or *nullify*. We can then accept the *alternative* to the null hypothesis. It must be understood that we can never *disprove* the null hypothesis or prove the alternative. We can only say that either the null hypothesis is untrue or a very improbable event has occurred.

EXAMPLE 14.1. A coin is tossed 100 times and it is found that it comes up tails 60 times. What null hypothesis should be used to determine whether the coin is balanced? (In other words, that a chance of a head or tail is 1/2.)
*Solution.* We assume that the coin is balanced and then examine the probability of 60 tails occurring in 100 tosses.

**14.3. Level of Significance.** It is customary to use a level of significance of 5% or 1%. With a level of 5%, for example, we accept an hypothesis if our test shows a 95% chance of its being correct. In this case, the test will show that the chance that the null hypothesis is correct is 5% or less. Similarly, with a 1% level of significance, the percentages are 99% and 1%.

**14.4. Type I and Type II Errors.** Pure chance will occasionally give test results which cause us to reject an hypothesis when it should be accepted. This is called a Type I Error. If, however, we accept an hypothesis which should be rejected, this is a Type II Error. Sometimes one type of error is more serious than the other, and where necessary, tests must be designed to reflect this.

**14.5. One and Two-Tailed Tests.** We are usually concerned with testing the significance of the departure of an observed value from the value indicated by our hypothesis. This involves finding the area under the normal curve which represents a departure either above or below the assumed figure. In other words, we are concerned with measuring the *two tails* of the normal curve. In some cases, we are interested in a difference in one direction only. In this case, we use a one-tailed test.

EXAMPLE 14.2. Give an example where a one-tailed test would be used.
*Solution.* In Problem 13.16, a poll reported 54% of the electorate voting for candidate $A$ and we calculated the probability, under certain assumptions, that he would be elected. If we wished to test the hypothesis that candidate $A$ would be elected, we would be concerned with the observed value (54%) being more than 4 percentage points above the true value, but not with it being more than 4 percentage points below the true value.

**14.6. Significance Levels.** These are determined by the areas under the normal curve.

| Significance Level | Critical Value ($z$) | |
|---|---|---|
| | Two-Tailed Test | One-Tailed Test |
| 1% | 2.58 | 2.33 |
| 5% | 1.96 | 1.64 |
| 10% | 1.64 | 1.28 |

Other values can be calculated from Table 7.1.

It will be noted that the value of $z$ for $x$% significance under a two-tailed test is equal to that for $(100 - x)$% confidence in the table in the previous chapter, since both are calculated from the normal curve. The value of $z$ for $x\%$ significance under a one-tailed test is equal to that for $2x\%$ significance under a two-tailed test. For small samples, the values of $z$ must be adjusted as described in Chapter 12. (Section 12.10, Student's $t$-distribution.)

**14.7. Step-by-Step Test Procedures.** The procedures in testing an hypothesis are as follows—

1. Decide on a theoretical model for the population.
2. Decide on the statistical hypothesis to be tested.
3. Decide on a statistic to be calculated from the observations for the purpose of testing.
4. Calculate the significance level of the test.
5. Make a subjective conclusion based on the test.

EXAMPLE 14.3. Use the data of Example 14.1, where a coin came up tails 60 times in a 100 tosses, to determine if the coin was balanced.

*Solution.* The theoretical model we select is that the probability $p$ of a tail on any throw is the same as the probability of a tail on any other throw. The statistical hypothesis to be tested is that the probability $p$ is equal to 1/2, that is, that the coin is balanced. The statistic to be calculated is the probability of 60 or more tails in 100 tosses.

On our hypothesis, the mean number of tails should be 50 with a standard deviation of $\sqrt{100 \times .5 \times .5} = 5$. Since the normal curve is continuous, and we use it as an approximation to a discrete distribution, 60 or more tails must be interpreted as $59\frac{1}{2}$ or more. We have to calculate the probability of a departure from the mean of $59.5 - 50 = 9.5$ or more.

This is $1.9\sigma$, which for a one-tailed test has a significance level of 2.9%. Using a two-tailed test, we would calculate the probability of (60 of more) or (40 or less) tails which would be 5.8%. We now come to the subjective conclusion that in the light of these low significance levels it is unlikely that the coin is balanced.

EXAMPLE 14.4. What is the null hypothesis and the alternative hypothesis in the above example?

*Solution.* The null hypothesis is $p = 1/2$. The alternative hypothesis is $p > 1/2$. On the basis of the one-tailed test, the null hypothesis is rejected at the 3% significance level.

EXAMPLE 14.5. If we accept the hypothesis that the coin is balanced and it is in fact unbalanced, what type of error is involved?
*Solution.* This is a Type II error.

**14.8. Sample Differences.** If we have two samples, we can test whether they come from the same population by using the null hypothesis that the two populations are the same. Similar tests can be used with proportions.

EXAMPLE 14.6. An I.Q. test is applied to two groups of 100 students, all the same age. If the mean I.Q.'s are 105 and 110 for the two samples, and the standard deviation of I.Q.'s is assumed to be 20, test the assumption that the two groups can be assumed to come from the same population.
*Solution.* The difference in the means is 5 and the standard deviation of the difference is (by Formula 13.3)

$$\sqrt{\frac{\sigma^2}{N_1} + \frac{\sigma^2}{N_2}} = \sqrt{\frac{20^2}{100} + \frac{20^2}{100}} = \sqrt{\frac{800}{100}} = 2.8$$

The difference is

$$\frac{5}{2.8}\sigma = 1.8\sigma$$

This is a two-tailed test and the significance level is 7%. It is not unreasonable to assume that the two groups came from the same population.

**14.9. Chi-Square Test.** This test is used to determine the significance of the differences between observed and expected frequencies, where the expected frequencies are based on some hypothesis. In this test, the statistic calculated is

$$\chi^2 = \sum \frac{(o_i - e_i)^2}{e_i} \tag{14.1}$$

where the summation is made over the events for which observed and expected statistics are available. For event $i$, $o_i$ is the observed frequency and $e_i$ the expected frequency. $\chi$ is the Greek letter chi, and the test is called the chi-square test.

Values of $\chi^2$ are tabulated in statistical tables for various *de-*

*grees of freedom.* (See Chapter 12, Section 12.10.) Some typical values are given below.

**Values of $\chi^2$**

| **Degrees of Freedom** | **Significance Level** .20 | .10 | .05 | .01 |
|---|---|---|---|---|
| 1 | 1.6 | 2.7 | 3.8 | 6.6 |
| 2 | 3.2 | 4.6 | 6.0 | 9.2 |
| 3 | 4.6 | 6.3 | 7.8 | 11.3 |
| 4 | 6.0 | 7.8 | 9.5 | 13.3 |
| 5 | 7.3 | 9.2 | 11.1 | 15.1 |
| 10 | 13.4 | 16.0 | 18.3 | 23.2 |
| 20 | 25.0 | 28.4 | 31.4 | 37.6 |

Where an $h \times k$ table is involved, the number of degrees of freedom is

$$\nu = (h - 1)(k - 1) - m$$

where $m$ is the number of population parameters which have to be estimated for the sample statistics.

For discrete data, a correction known as *Yates' correction* should be used which changes Formula 14.1 to

$$\chi^2 = \sum \frac{(\,|\,o_i - e_i\,|\, - .5)^2}{e_i} \tag{14.2}$$

where $|\,o_i - e_i\,|$ is the positive value of $o_i - e_i$.

EXAMPLE 14.7. In a certain mortality investigation, the observed number of deaths and the expected number of deaths according to a certain mortality table were—

| **Age Group** | **Actual Deaths** | **Expected Deaths** |
|---|---|---|
| 20– | 58 | 47 |
| 30– | 78 | 80 |
| 40– | 96 | 117 |
| 50– | 101 | 115 |
| 60– | 46 | 36 |

Can the lives in the actual mortality investigation be considered as a sample from the population of the lives used to prepare the mortality table?

*Solution.*

| Age Group | Actual Deaths | Expected Deaths | A − E | (A − E)$^2$ | $\frac{(A-E)^2}{E}$ |
|---|---|---|---|---|---|
| 20– | 58 | 47 | +11 | 121 | 2.57 |
| 30– | 78 | 80 | − 2 | 4 | .05 |
| 40– | 96 | 117 | −21 | 441 | 3.77 |
| 50– | 101 | 115 | −14 | 196 | 1.70 |
| 60– | 46 | 36 | +10 | 100 | 2.78 |
| | 379 | 395 | −16 | | $\chi^2$ = 10.87 |

Since the expected deaths were taken from a separate mortality table, no population parameters have been calculated and the number of degrees of freedom is 5. $\chi^2$ at the 10% significance level is 9.2, and at the 5% significance level it is 11.1.

Hence, the hypothesis can be accepted at the 10% significance level, but not at the 5% significance level.

EXAMPLE 14.8. How would the above results be affected by the Yates correction?

*Solution.* Since the number of actual deaths is discrete, $(A - E)$ should be corrected as follows—

| Age Group | $\lvert A-E \rvert$ | $\lvert A-E \rvert -.5$ | $(\lvert A-E \rvert -.5)^2$ | $\frac{(\lvert A-E \rvert -.5)^2}{E}$ |
|---|---|---|---|---|
| 20– | 11 | 10.5 | 110 | 2.34 |
| 30– | 2 | 1.5 | 2 | .03 |
| 40– | 21 | 20.5 | 420 | 3.59 |
| 50– | 14 | 13.5 | 182 | 1.58 |
| 60– | 10 | 9.5 | 90 | 2.50 |
| | | | | 10.04 |

$$\chi^2 = 10.04$$

The conclusion is unaltered.

**14.10. Computing $\chi^2$ in a 2 × 2 Contingency Table.** The observed frequencies in a study which involves two classifications of each of two variables can be set out in a contingency table as follows—

| | I | II | Total |
|---|---|---|---|
| **A** | $a_1$ | $a_2$ | $N_A$ |
| **B** | $b_1$ | $b_2$ | $N_B$ |
| **Total** | $N_1$ | $N_2$ | $N$ |

If there is no correlation between the two variables, we shall expect the $N_1$ values in I to be distributed between $A$ and $B$ in the same proportion as the total. That is, we shall expect $a_1$ to be equal to

$$\frac{N_A}{N} \times N_1$$

and similarly for the other values. From this and the three other similar equations we can show that

$$\chi^2 = \frac{N(a_1 b_2 - a_2 b_1)^2}{N_1 N_2 N_A N_B} \tag{14.3}$$

With Yates' correction this becomes

$$\chi^2 = \frac{N(\,|\,a_1 b_2 - a_2 b_1\,| - \tfrac{1}{2}N)^2}{N_1 N_2 N_A N_B}$$

Example 14.9. A company using a door-to-door sales procedure is testing a new sales approach, and has the following results on a comparative test under otherwise identical conditions.

| | **Sales** | **No Sales** | **Total Visits** |
|---|---|---|---|
| **Old Approach** | 84 | 116 | 200 |
| **New Approach** | 98 | 102 | 200 |
| **Total** | 182 | 218 | 400 |

Use the $\chi^2$ test to determine the significance of the observed difference.

*Solution.* The expected frequencies from the combined data are

$$a_1 = b_1 = 91$$

$$a_2 = b_2 = 109$$

Calculating $\chi^2$ directly,

$$\chi^2 = \frac{(84 - 91)^2}{91} + \frac{(98 - 91)^2}{91} + \frac{(116 - 109)^2}{109} + \frac{(102 - 109)^2}{109}$$

$$= .54 + .54 + .45 + .45$$

$$= 1.98$$

Alternatively, using Formula 14.3,

$$\chi^2 = \frac{400[(84 \times 102) - (98 \times 116)]^2}{182 \times 218 \times 200 \times 200} = 1.98, \text{ as before.}$$

$$\nu = (h - 1)(k - 1) = (2 - 1)(2 - 1) = 1$$

so that we have one degree of freedom.

Since $1.6 < 1.98 < 2.7$ we conclude that the new approach has significance at the 20% level, but not at the 10% level. It seems probable that the new approach has not significantly improved the results, but further tests are desirable.

**14.11. 2 × 3 and Larger Contingency Tables.** From Formula 14.1

$$\chi^2 = \sum \frac{(o - e)^2}{e} = \sum \frac{o^2}{e} - 2\sum \frac{oe}{e} + \sum \frac{e^2}{e}$$

$$= \sum \frac{o^2}{e} - 2\Sigma o + \Sigma e$$

In any contingency table, $\Sigma o = \Sigma e = N$
Hence,

$$\chi^2 = \sum \left(\frac{o^2}{e}\right) - N \qquad (14.4)$$

**14.12. Coefficient of Contingency and Correlation of Attributes.** The *coefficient of association or contingency* is a useful measure of association between two or more attributes. It is defined as

$$C = \sqrt{\frac{\chi^2}{\chi^2 + N}}$$

The larger the value of $C$, the greater the association. The maximum value of $C$ is

$$\sqrt{\frac{k - 1}{k}}$$

where $k$ is the number of rows and columns in the contingency table. $C$ is always less than 1.

Another useful measure is the *correlation of attributes* which is defined as

$$r = \sqrt{\frac{\chi^2}{N(k - 1)}}$$

The value is always less than 1.

EXAMPLE 14.10. From Example 14.9, calculate the coefficient of contingency and the correlation of attributes.
*Solution.* The coefficient of contingency is

$$C = \sqrt{\frac{\chi^2}{\chi^2 + N}}$$

where

$$\chi^2 = 1.98 \quad \text{and} \quad N = 400$$

giving

$$C = \sqrt{\frac{1.98}{401.98}} = \sqrt{.0049} = .07$$

The correlation of attributes is

$$r = \sqrt{\frac{1.98}{400(2 - 1)}} = .07$$

## Problems

**Problem 14.1.** A die is to be rolled 216 times and the number of times a six comes up counted. It is decided to accept the hypothesis that the die is true if the number of sixes is any number from 31 to 41 and to reject this hypothesis if the number of sixes is 30 or less or 42 or more. List the step by step test procedures. Why is this test procedure faulty?
**Problem 14.2.** In Problem 14.1, what is the probability of rejecting the hypothesis that the frequency of sixes is 1/6 when it is actually correct.
**Problem 14.3.** In Problem 14.1, what range of values should be chosen to avoid rejecting the hypothesis that the frequency of sixes is 1/6 at the 5% significance level?
**Problem 14.4.** How many times should the die in Problem 14.1 be rolled if a 10% departure from the expected number of sixes (on the assumption of a frequency of 1/6) is to correspond to the 10% significance level?
**Problem 14.5.** To test whether a student has any knowledge of a subject, he is set 6 true or false questions and told to guess if he

does not know the answer. The hypothesis is made that if 4 or more answers are right the student has some knowledge of the subject.

What is the probability that a student who guesses the answer to all 6 questions makes this score?

**Problem 14.6.** There are 100 questions in a multiple choice (5 answer) test, so as to discourage guessing, the score is calculated by the following formula

$$R - \frac{1}{4}W$$

where $R$ equals the number of correct answers and $W$ the number of wrong answers. The passing score is 50. A student knows the answer to 40 questions. Of the remaining 60, he knows that in 20 only two out of the 5 possible answers are correct. He has no means of guessing the correct answer to the remaining 40 questions. If he uses a random method of guessing the 20 questions, where he can exclude answers he knows to be wrong, what is the probability that he will pass the examination?

**Problem 14.7.** 300 balls are drawn at random from a bag with replacement after each drawing. The hypothesis is made that there are two black balls for each red ball, and this is accepted if the number of red balls drawn is in the range 90 to 110. Calculate the probability of a Type I error. What is the significance level of this test?

**Problem 14.8.** If, in Problem 14.7, suppose there were in fact three black balls for every two red balls in the bag. What is the probability of accepting the two black balls to one red ball hypothesis when it should be rejected (Type II error)?

**Problem 14.9.** Illustrate the results of Problems 14.7 and 14.8 graphically.

**Problem 14.10.** If a two-tailed test has a critical value of $2 \times \sigma$, what is the significance level? What is the significance level if it is a one-tailed test?

**Problem 14.11.** Students in two classes take the same test. The results are

| Class | No. of Students | Mean Score | Standard Deviation |
|---|---|---|---|
| *A* | 50 | 58 | 10 |
| *B* | 40 | 62 | 8 |

At what level of significance are the results different?

**Problem 14.12.** If in the previous study, the figures had referred to two random samples of the same class, sub-division $B$ having been given an extra hour a week of study, and we wanted to know if the additional study time had significantly improved the results, what difference would be made in the test?

**Problem 14.13.** Two samples have identical means. Can we conclude they must come from the same population?

**Problem 14.14.** A die is rolled 360 times and the results are tabulated.

| **Face** | 1 | 2 | 3 | 4 | 5 | 6 |
|---|---|---|---|---|---|---|
| **Frequency** | 46 | 65 | 71 | 70 | 56 | 52 |

Use the chi-square test to determine if the die is true.

**Problem 14.15.** If instead of a manufactured die, in the previous problem, a home-made die had been used, how would the answer be altered?

**Problem 14.16.** If the die in Problem 14.14 were rolled twice as many times and the identical frequency distribution resulted, would this increase, leave unchanged, or decrease the significance level of the test?

**Problem 14.17.** Prove Formula 14.3 algebraically.

**Problem 14.18.** Use the chi-square method on the data in Example 14.3.

**Problem 14.19.** Calculated chi-square for the following contingency table. (Ignore the Yates correction.)

| | Hair Color | | | |
|---|---|---|---|---|
| | **Fair** | **Dark** | **Redhead** | **Totals** |
| **Boys** | 40 | 50 | 10 | 100 |
| **Girls** | 60 | 30 | 10 | 100 |
| | 100 | 80 | 20 | 200 |

What is the significance level of these differences?

**Problem 14.20.** What is the coefficient of contingency for the data in the previous problem?

## Solutions

**Problem 14.1.** The step-by-step test procedures are—

1. The theoretical model is that the probability of throwing a six on any roll is the same.
2. The statistical hypothesis to be tested is that the chance of a six is 1/6 on any roll.

3. Calculate the probability that the number of sixes lie outside the range from 31 to 41 in 216 rolls.
4. Calculate the significance level of this test.
5. Make a subjective conclusion based on the test.

The test procedure is faulty because it will test only the proper frequency of the distribution of sixes. It applies no test to the other faces of the die. The die might be so weighted that the frequencies for the various faces were—

| **Face** | 1 | 2 | 3 | 4 | 5 | 6 |
|---|---|---|---|---|---|---|
| **Frequency** | 1/8 | 1/4 | 1/8 | 1/12 | 1/4 | 1/6 |

**Problem 14.2.** $Np = 36$ and $Nq = 180$ and the normal approximation to the binomial may be used.

$$\mu = Np = 36$$
$$\sigma = \sqrt{Npq} = \sqrt{30} = 5.48$$

On a continuous scale, the test applied is that the range 30.5 to 41.5 equals $\mu \pm \sigma$. From Table 7.1 the area is .68. Hence, the probability that the hypothesis will be rejected when it is in fact true is .32.

**Problem 14.3.** For a two-tailed test, the critical value for a 5% significance level is 1.96 (see Section 14.6).

$$1.96 \times 5.48 = 10.7$$

The range of values should be 26 to 46 inclusive.

**Problem 14.4.** For a 10% departure, the range is

$$Np \pm \frac{1}{10} Np$$

For a two-tailed test, the critical value for 10% significance level is 1.64. Therefore

$$1.64\sigma = \frac{1}{10} Np$$

$$1.64\sqrt{Npq} = \frac{1}{10} Np$$

$$2.69N \times \frac{1}{6} \times \frac{5}{6} = \frac{1}{100} N^2 \left(\frac{1}{6}\right)^2$$

$$N = 2.69 \times 5 \times 100 = 1345$$

The number of required rolls is approximately 1,350.

**Problem 14.5.** For a student who guesses, the probability of a correct reply to an individual question is 1/2. The probability of 4 or more questions right is

$$\frac{6.5}{1.2}\left(\frac{1}{2}\right)^2\left(\frac{1}{2}\right)^4 + \frac{6}{1}\left(\frac{1}{2}\right)\left(\frac{1}{2}\right)^5 + \left(\frac{1}{2}\right)^6$$

$$= \frac{15 + 6 + 1}{64} = \frac{22}{64} = .34$$

Hence, the probability of rejecting the assumption that the student is guessing when this assumption in fact is correct is .34. (A Type I error.)

**Problem 14.6.** Of the twenty questions he can guess at, he gets 1 for a correct answer and $-1/4$ for an incorrect answer. In order to pass, he must guess 12 or more correctly, since

$$12 - \frac{(20 - 12)}{4} = 10$$

$Np = 10$ and $Nq = 10$ so that the binomial approximation may be used.

$$\sqrt{Npq} = \sqrt{20 \times 1/4} = \sqrt{5} = 2.24$$

$$\frac{(2 - .5)}{2.24} = .67$$

The area under the normal curve between $.67\sigma$ and $\infty$ is .25. Hence, the probability he will pass is .25. The same result is obtained by summing the terms of the binomial expansion.

**Problem 14.7.** For a Type I error, the hypothesis is rejected when it should be accepted.

$$p = 1/3 \qquad Np = 100 \qquad Nq = 200$$

$$\sqrt{Npq} = \sqrt{\frac{200}{3}} = \sqrt{66.7} = 8.17$$

The accepted range is $Np \pm 10.5$ or

$$\frac{10.5}{8.17}\sigma = 1.285\sigma$$

The area of the normal curve outside this range is $[1 - (2 \times .4)] = .2$. The probability of a Type I error is .2. The significance level of the test is 20%.

**Problem 14.8.** For a Type II error, the hypothesis is accepted when it should be rejected. This means that the test gives a value in the range 90 to 110.

$$p = 2/5 \qquad Np = 120 \qquad Nq = 180$$

$$\sqrt{Npq} = \sqrt{120 \times 3/5} = \sqrt{72} = 8.49$$

$$(120 - 110.5) = 9.5 = \frac{9.5}{8.49}\,\sigma \quad \text{or} \quad 1.12\sigma$$

The probability of a Type II error is

$$1 - .37 - .5 = .13$$

**Problem 14.9.**

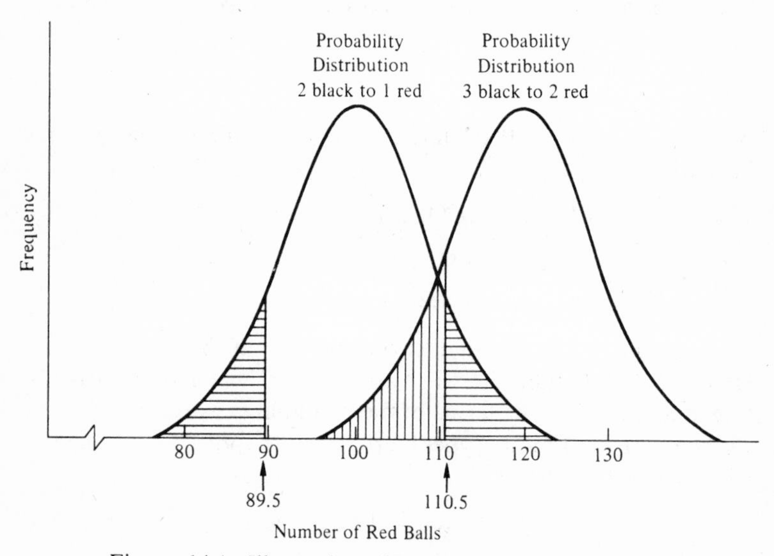


Figure 14.1. Illustration of Type I and Type II Errors.

The horizontal shading represents the answer to Problem 14.7, and the vertical shading represents the answer to Problem 14.8. Note that in one case, two tails of the curve are involved, and the other case, only one tail.

**Problem 14.10.** (1) 4.6% (2) 2.3%

**Problem 14.11.** The null hypothesis is that both classes came from the same population. The closest approximation to the standard deviation of the population of the two classes combined will be approximately 9.

$$\text{Difference in mean} = 62 - 58 = 4$$

The standard deviation of the difference is

$$\sqrt{\frac{q^2}{N_1} + \frac{q^2}{N_2}} = \sqrt{3.65} = 1.91$$

$$\text{critical value} = \frac{4}{1.91} = 2.09$$

For a two-tailed test, this gives a significance level of 4%.

**Problem 14.12.** In this case, a one-tailed test should be used and the significance level would be 2%.

**Problem 14.13.** First, a Type II error may be involved and the hypothesis may be accepted when it should be rejected. Second, while the means of the two samples may be the same, the standard deviation or the skewness may be so markedly different that it is improbable that the two samples come from the same population.

**Problem 14.14.**

| **Face** | 1 | 2 | 3 | 4 | 5 | 6 |
|---|---|---|---|---|---|---|
| **Observed Frequency** | 46 | 65 | 71 | 70 | 56 | 52 |
| **Expected Frequency** | 60 | 60 | 60 | 60 | 60 | 60 |
| **o − e** | −14 | +5 | +11 | +10 | −4 | −8 |
| **\|o − e\| − 1/2** | 13.5 | 4.5 | 10.5 | 9.5 | 3.5 | 7.5 |
| **[\|o − e\| − 1/2]²** | 182.3 | 20.3 | 110.3 | 90.3 | 12.3 | 56.3 |

Since the expected value is the same for each face we may sum $[|o - e| - 1/2]^2$ before dividing by $e$

$$\chi^2 = \sum \frac{[|o - e| - 1/2]^2}{e} = \frac{471.5}{60} = 7.86$$

Since the total observed and expected frequencies are the same, the number of restraints is 1, and the number of degrees of freedom is 5. The observed differences are significant at the 20% significance level but not at the 10% significance level. Since most die are true, we would not assume from this result that the die was untrue without further testing.

**Problem 14.15.** If the maker was skilled at cutting a die, we would probably come to the same conclusion. If however, the maker was poorly skilled we might well conclude the die was untrue. This illustrates the importance of the subjective conclusion which is the last step in decision theory (see Section 14.7).

**Problem 14.16.** The values of $(o - e)$ would be approximately doubled (the Yates correction makes the doubling not exact) and so would $e$. Since the numerators of the terms in $\chi^2$ are squared but not the denominators, the value of $\chi^2$ would be larger and consequently the significance level would be a smaller figure. $\chi^2$ becomes 16.5 and the die is untrue at the 1% significance level.

**Problem 14.17.** For the left top square, the actual value is $a_1$ and the expected value is

$$\frac{N_A}{N} \times N_1$$

$$(o_1 - e_1) = a_1 - \frac{(a_1 + a_2)(a_1 + b_1)}{N}$$

$$= \frac{a_1(a_1 + a_2 + b_1 + b_2) - (a_1 + a_2)(a_1 + b_1)}{N}$$

$$= \frac{a_1 b_2 - a_2 b_1}{N}$$

$$\frac{(o_1 - e_1)^2}{e_1} = \frac{(a_1 b_2 - a_2 b_1)^2}{N N_A N_1}$$

$$\chi^2 = \sum \frac{(o - e)^2}{e} = \frac{1}{N}(a_1 b_2 - a_2 b_1)^2 \cdot \left(\frac{1}{N_A N_1} + \frac{1}{N_A N_2} + \frac{1}{N_B N_1} + \frac{1}{N_B N_2}\right)$$

$$= \frac{1}{N}(a_1 b_2 - a_2 b_2)^2 \cdot \frac{N_B N_2 + N_B N_1 + N_A N_2 + N_A N_1}{N_1 N_2 N_A N_B}$$

$$= \frac{1}{N}(a_1 b_2 - a_2 b_1)^2 \frac{N_B N + N_A N}{N_1 N_2 N_A N_B}$$

$$= \frac{N(a_1 b_2 - a_2 b_1)^2}{N_1 N_2 N_A N_B}$$

**Problem 14.18.**

$$o_1 = 60 \qquad o_2 = 40 \qquad e_1 = e_2 = 50$$

$$\chi^2 = \frac{(|60 - 50| - .5)^2}{50} + \frac{(|40 - 50| - .5)^2}{50}$$

$$= 3.61$$

The degrees of freedom are $(2 - 1) \times (2 - 1) = 1$ and hence the test is significant at just about the 5% significance level (Section 14.9). This agrees with the result obtained in Example 14.3 using the two-tailed test. Note that the square of 1.9, the standard score, in the solution to Example 14.3, is equal to $\chi^2$. This is always true when only two categories are involved.

**Problem 14.19.** The expected values are

| | Hair Color | | | |
|---|---|---|---|---|
| | **Fair** | **Dark** | **Redhead** | **Totals** |
| **Boys** | 50 | 40 | 10 | 100 |
| **Girls** | 50 | 40 | 10 | 100 |
| **Totals** | 100 | 80 | 20 | 200 |

Calculating $\chi^2$ directly—

$$\frac{(40 - 50)^2}{50} + \frac{(50 - 40)^2}{40} + \frac{(10 - 10)^2}{10} + \frac{(60 - 50)^2}{50} + \frac{(30 - 40)^2}{40} + \frac{(10 - 10)^2}{10}$$

$$= 9.0$$

or using Formula 14.4

$$\frac{40^2}{50} + \frac{50^2}{40} + \frac{10^2}{10} + \frac{60^2}{50} + \frac{30^2}{40} + \frac{10^2}{10} - 200$$

$$= 9.0$$

The number of degrees of freedom is $(3 - 1) \times (2 - 1) = 2$, indicating a significance level of 1%.

**Problem 14.20.** The coefficient of contingency is

$$\sqrt{\frac{9}{9 + 200}} = \sqrt{\frac{9}{209}} = .21$$

# Index